DEFENSE, SECURITY AND STRATEGY SERIES

WAR IN AFGHANISTAN: STRATEGY, MILITARY OPERATIONS AND CONGRESSIONAL ISSUES

DEFENSE, SECURITY AND STRATEGY SERIES

Military Satellites: Issues, Goals and Challenges
Abel Chirila (Editor)
2009. ISBN: 978-1-60741-238-0

Evaluating Military Compensation
Jaime G. Duenas (Editor)
2009. ISBN: 978-1-60741-476-6

The Army's Future Combat System Program
Christian N. Feliciano (Editor)
2009. ISBN: 978-1-60741-262-5

Strategizing Resilience and Reducing Vulnerability
Peter R. J. Trim and Jack Caravelli (Editors)
2009. ISBN: 978-1-60741-693-7

Counterinsurgency and the Armed Forces
Laure Paquette
2009. ISBN: 978-1-60741-763-7

U.S. Nuclear Stockpile: Maintenance and Replacement of Warheads
Robin A. Kraemer (Editor)
2010. ISBN: 978-1-60741-483-4

Security in Iraq
James L. Jones, Jennifer K. Elsea, and Nina M. Serafino (Editors)
2010. ISBN: 978-1-60692-127-2

War in Afghanistan: Strategy, Military Operations and Congressional Issues
Easton H. Ussery (Editor)
2010. ISBN: 978-1-60741-579-4

DEFENSE, SECURITY AND STRATEGY SERIES

WAR IN AFGHANISTAN: STRATEGY, MILITARY OPERATIONS AND CONGRESSIONAL ISSUES

Easton H. Ussery
Editor

Nova Science Publishers, Inc.
New York

For permission to use material from this book please contact us:
Telephone 631-231-7269; Fax 631-231-8175
Web Site: http://www.novapublishers.com

LIBRARY OF CONGRESS CATALOGING-IN-PUBLICATION DATA

War in Afghanistan : strategy, military operations, and congressional issues / editor, Easton H. Ussery.
p. cm.
Includes bibliographical references and index.
ISBN 978-1-60741-579-4 (hardcover : alk. paper)
1. Afghan War, 2001- 2. North Atlantic Treaty Organization--Afghanistan. 3. United States. Army--History--Afghan War, 2001- 4. Strategy. 5. Command of troops. 6. United States. Congress. 7. Civil-military relations--United States. I. Ussery, Easton H.
DS371.412.W36 2009
958.104'7--dc22
2009030339

Published by Nova Science Publishers, Inc. ✢ New York

TABLE OF CONTENTS

PREFACE

This book is an overview of the war in Afghanistan; strategy, military operations and issues for Congress. As the 111th Congress convenes and the Obama administration assumes office, the war in Afghanistan has entered its eighth year. With a deteriorating security situation and no comprehensive political outcome yet in sight, most observers view the war in Afghanistan as open-ended. By early 2009, a growing number of members of Congress, Administration officials, and outside experts had concluded that the effort - often called "America's other war" - required greater national attention. This book defines "unity of command" and describes how the United States has tried to adhere to this principle since 1914. The evolution of the command structure in Afghanistan since 2001 is also explored. The current command structure in Afghanistan is also analyzed.

Chapter 1 - The mission of the North Atlantic Treaty Organization (NATO) in Afghanistan is a test of the alliance's political will and military capabilities. Since the Washington Summit in 1999, the allies have sought to create a "new" NATO, capable of operating beyond the European theater to combat emerging threats such as terrorism and the proliferation of weapons of mass destruction. Afghanistan is NATO's first "out-of-area" mission beyond Europe. The purpose of the mission is the stabilization and reconstruction of Afghanistan. The mission has proven difficult because it must take place while combat operations against Taliban insurgents continue. Recent assessments of the situation in Afghanistan point to a rise in the overall level of violence due to increased Taliban military operations and an increase in terrorist-related activities.

U.N. Security Council resolutions govern NATO's responsibilities in Afghanistan. The NATO-led International Security Assistance Force (ISAF) faces formidable obstacles: shoring up a weak government in Kabul; using military capabilities in a distant country with rugged terrain; and rebuilding a country devastated by war and troubled by a resilient narcotics trade. NATO's mission statement lays out the essential elements of the task of stabilizing and rebuilding the country: train the Afghan army, police, and judiciary; support the government in counter-narcotics efforts; develop a market infrastructure; and suppress the Taliban.

ISAF has proceeded in four stages to extend its area of responsibility over the whole of Afghanistan. Although the allies agree on ISAF's mission, they continue to differ on how to accomplish it. Some allies do not want their forces to engage in counter-insurgency operations and have placed operational restrictions on their troops. The principal mechanism to rebuild Afghanistan are the Provincial Reconstruction Teams (PRT) composed of military and civilian officials and charged with extending the reach of the Afghan government by improving governance and rebuilding the economy. However, there are significant differences in how individual NATO governments run their PRTs. Until recently, only the United States wanted to engage directly in the destruction of poppy fields and drug facilities in countering the drug trade. Finally, continued turmoil in parts of Pakistan has complicated the effort to prevent the Taliban from infiltrating Afghanistan.

The 111th Congress will likely support the new Administration's policies in Afghanistan, but may seek a more coherent plan for integration, stabilization, and reconstruction operations. In 2008, Congress appropriated $31 billion for the conflict in Afghanistan (H.R. 2642/P.L. 110-252) Most observers predict that ISAF's efforts to stabilize Afghanistan will require a long-term commitment. U.S. leadership in Afghanistan may well affect NATO's cohesiveness, credibility and its future. This report will be updated as needed. See also CRS Report RL30588, *Afghanistan: Post-War Governance, Security, and U.S. Policy*, by Kenneth Katzman.

Chapter 2 - This Carlisle Paper discusses the traditional importance of unity of command in American doctrine and practice from World War I until now, and how this principle has been forsaken in the evolution of military command for Afghanistan. It examines the unprecedented departure from the principle of unity of command in Afghanistan in 2006, when Combined Forces Command-Afghanistan passed control of the ground fight to the International Security Assistance Force, and operations became split between several unified or "supreme" commanders in charge of U.S. Central Command, the North Atlantic Treaty Organization, and U.S. Special Operations Command. It argues for a renewal of understanding of the importance of unity of command, and recommends that the United States revert to the application of this principle by amending the Unified Command Plan to invest one "supreme commander" with responsibility for the current Operation ENDURING FREEDOM Joint Operations Area.

Chapter 3 - As the 111th Congress convenes and the Obama Administration assumes office, the war in Afghanistan has entered its eighth year. With a deteriorating security situation and no comprehensive political outcome yet in sight, most observers view the war in Afghanistan as open-ended. By early 2009, a growing number of Members of Congress, Administration officials, and outside experts had concluded that the effort – often called "America's other war" – required greater national attention. For the Government of the Islamic Republic of Afghanistan (GIRoA), the war is both a struggle for survival and an effort to establish sustainable security and stability. For the United States, the war in Afghanistan concerns the security of Afghanistan and the region, including denying safe haven to terrorists and helping ensure a stable regional security balance. For regional states, including India and Russia as well as Afghanistan's neighbors Pakistan and Iran, the war may have a powerful impact on the future balance of power and influence in the region. For individual members of the North Atlantic Treaty Organization (NATO), the war may be about defeating terrorist networks, ensuring regional stability, proving themselves as contributing NATO members, and/or demonstrating NATO's relevance in the 21st century.

Since 2001, the character of the war in Afghanistan has evolved from a violent struggle against al Qaeda and its Taliban supporters to a multi-faceted counterinsurgency (COIN) effort. In the aftermath of the terrorist attacks of September 11, 2001, the United States launched Operation Enduring Freedom (OEF) in order to end the ability of the Taliban regime to provide safe haven to al Qaeda and to put a stop to al Qaeda's use of the territory of Afghanistan as a base of operations for terrorist activities. In that first phase, U.S. and coalition forces, working with Afghan opposition forces, quickly removed the Taliban regime.

After the fall of the Taliban, the character of the war shifted to a multifaceted COIN effort aimed at smothering the diffuse insurgency by shoring up GIRoA efforts to provide security, governance, and economic development. The three areas are generally viewed as interdependent and mutually-reinforcing – security is a prerequisite for some governance and development efforts, and longer-term, sustainable security requires both functional governance and economic opportunity. As one pillar of the COIN campaign in Afghanistan, the Afghan and international military effort aims broadly at defeating the remnants of the Taliban and other insurgents, securing the population, and helping extend the reach of the Afghan government. The international military effort includes both the NATO-led International Security Assistance Force (ISAF), to which the United States contributes troops, and the separate U.S.-led OEF mission.

The U.S. Government faces key strategic and operational decisions about its further engagement in the war in Afghanistan. These may include clarifying U.S. national interests in Afghanistan and the region; defining clear strategic objectives based on those interests; determining which diplomatic, economic, and military approaches to adopt, and what resources to commit to support those approaches; prioritizing "Afghanistan" versus other national security imperatives; and helping marshal a coordinated application of international efforts. Avenues available to Congress for exercising oversight include authorizing and appropriating funding for U.S. efforts in Afghanistan and the region; shaping policy through directive legislation; holding oversight hearings to assess policy execution; and adjusting Administration reporting requirements. This report provides analysis of current developments and future options concerning the war in Afghanistan. It will be updated as events warrant

In: War in Afghanistan: Strategy…
Editor: Easton H. Ussery, pp. 1-32
ISBN: 978-1-60741-579-4

Chapter 1

NATO IN AFGHANISTAN: A TEST OF THE TRANSATLANTIC ALLIANCE

Vincent Morelli[1] and Paul Belkin[2]
[1] Section Research Manager
[2] Analyst in European Affairs

SUMMARY

The mission of the North Atlantic Treaty Organization (NATO) in Afghanistan is a test of the alliance's political will and military capabilities. Since the Washington Summit in 1999, the allies have sought to create a "new" NATO, capable of operating beyond the European theater to combat emerging threats such as terrorism and the proliferation of weapons of mass destruction. Afghanistan is NATO's first "out-of-area" mission beyond Europe. The purpose of the mission is the stabilization and reconstruction of Afghanistan. The mission has proven difficult because it must take place while combat operations against Taliban insurgents continue. Recent assessments of the situation in Afghanistan point to a rise in the overall level of violence due to increased Taliban military operations and an increase in terrorist-related activities.

U.N. Security Council resolutions govern NATO's responsibilities in Afghanistan. The NATO-led International Security Assistance Force (ISAF) faces formidable obstacles: shoring up a weak government in Kabul; using military capabilities in a distant country with rugged terrain; and rebuilding a country devastated by war and troubled by a resilient narcotics trade. NATO's mission statement lays out the essential elements of the task of stabilizing and rebuilding the country: train the Afghan army, police, and judiciary; support the government in counter-narcotics efforts; develop a market infrastructure; and suppress the Taliban.

ISAF has proceeded in four stages to extend its area of responsibility over the whole of Afghanistan. Although the allies agree on ISAF's mission, they continue to differ on how to accomplish it. Some allies do not want their forces to engage in counter-insurgency operations and have placed operational restrictions on their troops. The principal mechanism to rebuild Afghanistan are the Provincial Reconstruction Teams (PRT) composed of military and civilian officials and charged with extending the reach of the Afghan government by improving governance and rebuilding the economy. However, there are significant differences in how individual NATO governments run their PRTs. Until recently, only the United States wanted to engage directly in the destruction of poppy fields and drug facilities in countering the drug trade. Finally, continued turmoil in parts of Pakistan has complicated the effort to prevent the Taliban from infiltrating Afghanistan.

The 111th Congress will likely support the new Administration's policies in Afghanistan, but may seek a more coherent plan for integration, stabilization, and reconstruction operations. In 2008, Congress appropriated $31 billion for the conflict in Afghanistan (H.R. 2642/P.L. 110-252) Most observers predict that ISAF's efforts to stabilize Afghanistan will require a long-term commitment. U.S. leadership in Afghanistan may well affect NATO's cohesiveness, credibility and its future. This report will be updated as needed. See also CRS Report RL30588, *Afghanistan: Post-War Governance, Security, and U.S. Policy*, by Kenneth Katzman.

INTRODUCTION[1]

NATO's mission in Afghanistan is seen as a test of the allies' military capabilities and their political will to undertake a complex mission in a distant land and to sustain that commitment. Since the Washington NATO Summit in 1999, the allies have sought to create a "new" NATO, capable operating beyond the European theater to combat emerging threats such as terrorism and the proliferation of weapons of mass destruction (WMD). NATO is seeking to be "global" in its geographic reach and in the development of non-member partner states that can assist in achieving an agreed mission. This change in overall mission initially reflected a NATO consensus that the principal dangers to allied security lie distant from the treaty area and require new political tools and military capabilities to combat them.

Two military operations in Afghanistan seek to stabilize the country. *Operation Enduring Freedom* (OEF) is a counter-insurgency, combat operation led by the United States against the Taliban and al Qaeda insurgents, primarily in the eastern and southern parts of the country along the Pakistan border. OEF is not a NATO operation, although many coalition partners are NATO members. Approximately 20,000 troops operate in OEF, including approximately 18,000 U.S. forces.[2]

The second operation is the International Security Assistance Force (ISAF). ISAF was created by United Nations Security Council Resolution 1386 on December 20, 2001. Led by the United States, the ISAF mission was initially limited to Kabul. NATO took over command of ISAF in Afghanistan in August 2003. The Security Council passed the currently governing resolution, Res. 1776, on September 17, 2007. The Resolution calls upon NATO to disarm militias, reform the justice system, train a national police force and army, provide security for elections, and provide assistance to the effort to address the narcotics industry. By December 2008, ISAF had an estimated 51,000 troops from 40 countries, with the 26 NATO members providing the core of the force. The largest troop deployments come from the United States which has approximately 19,000 troops, the United Kingdom (8,700), Germany (3,600), Canada (2,700), Italy (2,300), France (2,700) the Netherlands (1,800), and Poland

(1,200)[3]. The NATO/ISAF mission in Afghanistan is led by U.S. General David McKiernan who assumed command in June 2008.

NATO's effort in Afghanistan is the alliance's first "out-of-area" mission beyond Europe. The purpose of the mission is the stabilization and reconstruction of Afghanistan. Although NATO has undertaken stabilization and reconstruction missions before, for example in Kosovo, the scope of the undertaking in Afghanistan is considerably more difficult. Taliban and al Qaeda insurgents are providing stiff resistence to the operation, Afghanistan has never had a well-functioning central government, the distance from Europe, and the country's terrain present daunting obstacles to both NATO manpower and equipment. Stabilization and reconstruction must therefore take place while combat operations, continue. And although the allies agree upon a general political objective of the ISAF mission, some have differing interpretations of how to achieve it.

The mission in Afghanistan is likely to be important for NATO's future. Several key NATO members, above all the United States, have insisted that the allies must generate the political will to counter the greatest threats to their security. Afghanistan provides a test of will against the concrete danger of international terrorism. Over the past several years, NATO governments have also repeatedly pledged to develop capabilities making their forces more expeditionary, flexible, and "deployable." The mission in Afghanistan provides a hard test of these capabilities.

NATO's mission in Afghanistan also tests U.S. leadership of the alliance. Some allies have questioned whether the U.S. commitment to the interests of the allies preserves the mutual sense of obligation that characterizes the alliance. The allies also believe that the United States, as a global power, must provide leadership and resources to counter the destabilizing influences upon Afghanistan of two neighboring states, Iran and particularly, Pakistan.

A highly respected German Marshall Fund poll has found a sharp decline in European public opinion towards U.S. leadership since 2002. In key European countries, the desirability of U.S. leadership in the world fell from 64% in 2002 to 36% in June 2008; the approval rating of former President Bush in these same countries fell from 38% in 2002 to 19% in 2008.[4] This decline complicated the effort of allied governments to sustain public support for the ISAF mission. Secretary of Defense Robert Gates gave credence to the political ramifications of the Iraq war when he said in February 2008, "I worry that for many Europeans the missions in Iraq and Afghanistan are confused.... Many of them...have a problem with our involvement in Iraq and project that to Afghanistan."[5] Whether Europe's public view toward U.S. world leadership will begin to change as the new U.S. Administration begins to govern will also be a test of how well NATO's mission in Afghanistan will be viewed.

The conflict in Afghanistan continues to present a growing challenge to NATO. Over the past two years, Taliban attacks have increased in scope and number, and Taliban fighters have adopted some of the tactics, such as roadside bombs and suicide attacks, used by insurgents in Iraq. In January 2008, a report issued by the Afghanistan Study Group, claimed that the year 2007 was the deadliest for American and international troops in Afghanistan since the 2001 invasion of Afghanistan.[6] However, in 2008 the violence continued to escalate with a reported 30% increase nationwide and an estimated 40% rise in attacks over 2007 in the U.S.-led eastern sector. As a result, 2008 has now recorded the most U.S. combat casualties (150) and Afghan deaths (5,500) of the war. In June 2008, a Taliban-led attack on a prison in Kandahar resulted in the release of several hundred Taliban inmates. On July 13, 2008, a Taliban attack on a joint U.S.-Afghan outpost along the eastern border with Pakistan resulted in the death of 9 U.S. troops. This attack was the deadliest against U.S. forces in Afghanistan since 2005. On August 20, 2008, French forces suffered their worse combat casualties when 10 soldiers were killed after an ambush of a patrol by Taliban forces.

In addition to increased insurgent activity, terrorist violence has increased since the beginning of 2008. In February 2008, a terrorist bomb killed over 70 civilians and police officers near Kandahar. In April 2008 an assassination attempt was carried out against President Karzai.

The north and west of Afghanistan, and Kabul in the east, are largely stable, but combat operations in the south and east against the Taliban and al Qaeda continue. Forces from the United States, Britain, Canada, and the Netherlands, which are deployed in the east and south, bear the brunt of the fighting. The inequity of burden-sharing in combat operations remains an important point of contention within the alliance, and is a factor in the domestic opposition to the conflict in states that contribute the most combat forces.

Turmoil in neighboring Pakistan has further complicated ISAF's mission.[7] The assassination of presidential candidate Benazir Bhutto in December 2007, possibly by Islamic extremists, led to increasing internal restiveness against President Pervez Musharraf, criticized by some NATO governments as unable or unwilling to stem Taliban movement across the Pakistan border into Afghanistan. Some experts see Pakistani and Afghan Taliban militants increasingly merging and pooling their efforts against governments in both countries; Al Qaeda is reportedly actively facilitating the Afghanistan insurgency. Since the resignation of Musharraf, the new government in Pakistan has dispatched military units to the border region and has authorized the army to conduct offensive operations against Taliban forces in the northern tribal areas. In October 2008, the Pakistan government announced that it would begin to arm anti-Taliban tribal militias in the northern region in an attempt to control Taliban activity.[8] With the inability of the Pakistani government to control the number of Taliban insurgents who use Pakistan as a sanctuary, the United States has stepped up its use of missile attacks against suspected insurgent hideouts. This has caused a deterioration in U.S.-Pakistan relations to the point where there have been some shooting incidents between Pakistani forces and U.S. forces patrolling the Afghan border area. U.S. officials, in July 2008, confronted Pakistani officials with evidence that Pakistan's Inter-Services Intelligence agency (ISI) was actively helping Afghanistan militants, particularly the Haqqani faction.[9]

In another development, intelligence sources suggest that there has been an increase in the number of pro-al Qaeda foreign militants arriving in Pakistan from Iraq and other places in the Middle East. These sources believe these new arrivals continue to join Taliban fighters in Afghanistan.

The Karzai government in Afghanistan has come under international criticism, and its public support has diminished, due to corruption and an inability to improve living conditions. Some warlords continue to exert influence, and the narcotics industry remains an entrenched threat to the country's political health. The allies are not in full agreement on how to counter these problems, but allied officials say that they need a strong, competent, and reliable Afghan government to provide reasonable services to the population if NATO is to succeed.

The 110th Congress largely supported Administration policy in Afghanistan and it is expected that the 111th Congress will follow those same general lines of support. However, a number of congressional committees had called on the previous Administration to develop a more coherent plan to coordinate ISAF's stabilization and reconstruction efforts and it is expected that the same sentiments will be expressed to the new Obama Administration. In response, the Bush Administration led an effort before NATO's Bucharest summit in April 2008 to develop a "strategic vision" white-paper for Afghanistan that laid out a rationale for the mission that could be used to garner more public support for ISAF. The paper made four principal points: the allies promised a "long-term commitment" to Afghanistan; expressed support to improve the country's governance; pledged a "comprehensive approach" to bring civil and military efforts to effect stabilization; and promised increased engagement with Afghanistan's neighbors, "especially Pakistan."[10]

The paper represented some strides in bringing together allied views, but it also masked some important differences. It committed the allies to an indefinite period of time to stabilize Afghanistan, something that several allies had previously resisted. The paper, however, did not commit governments to pledge more forces; rather, the phrase "comprehensive approach" was seen by some observers as a euphemism for equating the importance of reconstruction and combat. Some governments believe that the military commitment remains paramount if security in the country is to improve so that reconstruction may proceed throughout Afghanistan. The paper also did not present a plan for engaging Pakistan or Iran; instead, the allies would continue to do so bilaterally, an approach that has not thus far yielded success in stemming the flow of arms or fighters into Afghanistan.[11]

This report follows the path of NATO's evolution in Afghanistan and addresses the current situation on the ground. The first section covers the initial two stages of ISAF's mission, and analyzes key issues in the mission: use of Provincial Reconstruction Teams to stabilize and rebuild the country; overcoming caveats placed by individual allies on the use of their forces; and managing the counter-narcotics effort. The next section of the report examines stages three and four of the ISAF mission which cover roughly the period December 2005 to the present. In this section, the debate to develop a refined mission statement and a new organizational structure is analyzed by looking at issues that are both political and military, such as securing more troops, the treatment of prisoners, and organization of command. By late 2006 as ISAF extended its responsibilities to cover all of Afghanistan, the allies began to realize that ISAF would require a greater combat capability than originally believed, and the mission would have to change. This adjustment in mission is discussed through the perspective of several key allies. The final sectionof the report assesses ISAF's progress to date.

Stages One and Two: Evolution of NATO in Afghanistan

Purpose of the Mission

The International Security Assistance Force (ISAF) was created by United Nations Security Council Resolution 1386 on December 20, 2001. Led by the United States, the ISAF mission was initially limited to Kabul. The United Nations, at the request of Afghan President Hamid Karzai, then asked for NATO's participation. NATO took over command of ISAF in Afghanistan in August 2003. The Security Council passed the most recent resolution, Res.

1776, on September 17, 2007. Like its predecessors, it calls upon NATO to disarm militias, reform the justice system, train a national police force and army, provide security for elections, and combat the narcotics industry. The resolution does not provide details of how NATO should accomplish these tasks; rather, the allies among themselves, in consultation with the Afghan government, have refined the resolution's provisions into active policy. The International Security Assistance Force includes troops from all 26 member states of the NATO alliance and has included troops from several non-NATO nations, such as Australia, New Zealand, Sweden, Jordan, and Azerbaijan.

Over time, NATO commanders laid out and implemented four stages designed to bring all of Afghanistan under NATO's operational responsibility. In Stage One in 2003-2004, NATO moved into the northern part of the country; French and German forces predominate in these areas. Stage Two began in May 2005, when NATO moved into western Afghanistan; Italian and Spanish forces are the core of the NATO force there. These sections of the country are relatively stable. Stage Three began in July 2006 when ISAF moved into the volatile southern region of Afghanistan, where U.S., British, Canadian, and Dutch forces predominate. Stage Four began in October 2006, when ISAF extended its geographic area of responsibility to include the entire country.

Issues

From the beginning of NATO's command of ISAF, political leaders and local commanders have had to deal with several significant issues which have influenced the implementation of the ISAF mission.

National Caveats

At the outset, NATO leaders faced considerable difficulty persuading some member states to contribute forces to ISAF. More importantly, however, a significant problem has been how some of those forces actually provided would operate once deployed. Many allies have committed forces to the NATO operation, then imposed restrictions—"national caveats"—on tasks those forces could undertake. It is reported that almost half the forces in ISAF have some form of caveats. National "caveats" or restrictions that allied governments, or their parliaments, place on the use of their forces, continue to trouble ISAF. While caveats in themselves do not generally prohibit the kinds of operations NATO forces can engage in, caveats do pose difficult problems for commanders, who seek maximum flexibility in utilizing troops under their command. Some governments' troops lack the appropriate equipment to function with other NATO forces. Some nations will not permit their troops to deploy to other parts of Afghanistan. Still others prohibit their troops from participating in combat operations unless in self-defense. NATO commanders have willingly accepted troops from some 40 governments but have had to shape the conduct of the mission to fit the capabilities of and caveats on those troops.

NATO commanders have long sought to minimize the number of caveats on forces dedicated to ISAF, an effort that has met with mixed success. In September 2006, former NATO SACEUR General James Jones expressed frustration at the limitations that some allies placed on their troops. "It's not enough," he said, "to simply provide forces if those forces have restrictions on them that limit them from being effective."[12]

At the NATO summit in Riga, Latvia in November 2006, some allied political leaders sought to reduce the caveats placed on forces in Afghanistan. The United States, Canada, Britain, and the Netherlands, which have forces in the highly unsettled areas of southern and eastern Afghanistan, appealed to other governments to release combat forces to assist them in moments of danger. Some progress has been made over time in persuading NATO members to adopt more flexible rules of engagement but those nations whose forces continue to bear a higher burden of risk continue to appeal to their partners for relief.

At the Bucharest summit in April 2008, NATO leaders again pledged to continue to work to remove the limitations placed on their troops. Some allies had singled out Germany for special criticism, given that Germany has a large contingent of over 3,000 troops most of which are deployed in what has been a relatively quiet area of northern Afghanistan. German troops reportedly patrol only in armored personnel carriers, and do not leave their bases at night.[13] This has led some to suggest that the implementation of excess force protection measures by the Germans has made their work, even in a safe area, far less effective. Former NATO SACEUR General Jones complained about German restrictions after he had specifically requested that Germany send some of its force in northern Afghanistan into the south to help combat Taliban activity, a request the German government initially refused. Since then, however, Germany has allowed some of its forces to respond in emergency situations.

The French government has somewhat reduced its caveats and agreed to allow its forces in Kabul and elsewhere come to the assistance of other NATO forces in an emergency. The Italian and Spanish governments have said that their force commanders in the field could make the decision to send forces to assist in an urgent situation. It remains unclear whether and when these commanders would have to request permission from their capitals to do so, a complicating factor that could delay a decision. Turkey, in contrast, has not changed its proscription against its forces' use in combat.

While there have been criticisms of NATO's role in Afghanistan, many NATO/ISAF forces do engage in offensive operations. Since 2006, NATO/ISAF combat forces have launched several operations, including Operation Medusa in 2006 aimed at ousting Taliban forces in Kandahar province. In 2007, NATO and Afghan forces retook the town of Musa Qala in Helmand province and conducted operations, Achilles and Silicon against Taliban forces. In 2008, in reaction to increased operations by the Taliban, NATO forces have increased the number of offensive operations they have engaged in.

The concern over the impact of national caveats has spread even beyond NATO itself. On July 9, 2008, the European Parliament debated and voted on a report on Afghanistan presented by its Committee on Foreign Affairs. One of the provisions in the report emphasized "that a major strengthening of political will and commitment is necessary, and that this should be followed up not only by a willingness to provide additional combat troops in the most difficult areas, unrestricted by national caveats..."[14]

Provincial Reconstruction Teams

Provincial Reconstruction Teams (PRTs) are civilian-military units of varying sizes designed to extend the authority of the central government into the countryside, provide security, and undertake projects (such as infrastructure development and the delivery of basic services) to boost the Afghan economy. Although some allied governments believe that poor governance, rather than the insurgency, is the principal problem impeding stabilization of the country, NATO officials describe the PRTs as the "leading edge" of the allies' effort to stabilize Afghanistan.

There are 26 ISAF-led PRTs in operation. Virtually all the PRTs, including those run by the United States, now operate under ISAF but with varying lead nations. Each PRT operated by the United States is composed of U.S. forces (50-100 U.S. military personnel); Defense Department civil affairs officers; representatives of USAID, State Department, and other agencies; and Afghan government (Interior Ministry) personnel. Most PRTs, including those run by partner forces, have personnel to train Afghan security forces. Many U.S. PRTs in restive regions are "colocated" with "forward operating bases" of 300-400 U.S. combat troops. U.S. funds support PRT reconstruction projects. According to U.S. officials in March 2008, 54 PRT development projects have been completed and 199 (valued at $20 million) are ongoing.

In August 2005, in preparation for the establishment of Regional Command South, Canada took over the key U.S.-led PRT in Kandahar. In May 2006, Britain took over the PRT at Lashkar Gah, capital of Helmand Province and the area of recent heavy fighting in the Fall of 2008. The Netherlands took over the PRT at Tarin Kowt, capital of Uruzgan Province. Germany (with Turkey and France) took over the PRTs and the leadership role in the north from Britain and the Netherlands when those countries deployed to the south.

Representing evolution of the PRT concept, Turkey opened a PRT in Wardak Province on November 25, 2006, to focus on providing health care, education, police training, and agricultural alternatives in that region. In March 2008, the Czech Republic established a PRT in Lowgar Province. South Korea has taken over the U.S.-run PRT at Bagram Air Base. There also has been a move to turn over the lead in the U.S.-run PRTs to civilians rather than military personnel, presumably State Department or USAID officials. That process began in early 2006 with the establishment of a civilian-led U.S.-run PRT in the Panjshir Valley.[15]

There is no established model for PRTs, and many are dominated by military forces, rather than civilian technicians. By most accounts, those serving in PRTs make an effort to move about surrounding territory, engage the local governments and citizens, and demonstrate that the international presence is bringing tangible results. Despite general support for PRTs, they receive mixed reviews and there have been criticisms of the overall PRT initiative. Some observers believe the PRTs operate without an overarching concept of operations, do not provide a common range of services, do not have a unified chain of command, and often do not coordinate with each other or exchange information on best practices.[16]

Another problem that has risen for PRTs in some areas is that civilian relief organizations do not want to be too closely associated with the military forces assigned to the PRTs because they feel their own security is endangered as well as their perceived neutrality. On September 10, 2009, U.S. Defense Secretary Robert Gates testifying before the House Armed Services Committee stated that "absent a broader international and interagency approach to the problems there... no amount of troops in no amount of time can ever achieve all the objectives we seek in Afghanistan." He went on to say that "Afghanistan doesn't just need more boots on the ground. It needs more trucks, teachers judges... foreign investment, alternative crops, sound governance, and rule of law. These are the keys to success in Afghanistan. No armed force, anywhere, no matter how good, can deliver these keys alone.[17]

Although U.S. and ISAF PRTs share the same mission there are reportedly considerable differences in structure. U.S. PRTs are composed of military personnel, civil affairs officers, representatives of the U.S. and other government agencies focused on reconstruction, and Afghan government personnel. Some observers believe U.S. PRTs are too heavily weighted with military personnel who lack the expertise to assist in developing important elements of the economy.[18] Others believe that there is a lack of qualified civilian personnel to accomplish key tasks. For instance, some claim that there is a critical shortage of U.S. agricultural specialists on the ground in Afghanistan.[19] The United States government controls the funds for its PRTs, in part to ensure that the money does not disappear through the hands of corrupt officials in the provinces or in Kabul, and that it goes directly to designated projects.

ISAF PRTs generally have fewer personnel as well as a different mix of military forces and civilian experts. Some U.S. officials believe that European-led PRTs are too hesitant in their engagement of the Afghan population. Some European-led PRTs are minimally funded, or provide little supervision of how their funds are managed and dispensed.[20] The Dutch, for instance, give their funding for PRT reconstruction activities directly to the Afghan central government, mainly through U.N. and World Bank channels. The Dutch argue that the Karzai government itself must undertake responsibility for planning and implementation of projects to rebuild the country. By contrast, the French have declined to lead a PRT and have questioned NATO's role in the PRTs.

In hearings before the 110th Congress, witnesses urged steps to strengthen the PRTs. Some witnesses argued that the Administration should increase funding for the State Department, AID, and the U.S. Department of Agriculture, three parts of the government able to provide needed expertise in the PRTs. Witnesses also repeatedly called for a model for ISAF PRTs that might provide guideposts to "best practices" to ensure a higher quality of assistance to the Afghan population.[21] The new Administration has begun to discuss a more enhanced role for the Department of State in tasks such as those required in Afghanistan and it is likely the 111th Congress will continue to press for a more coherent reconstruction effort.

Counter-Narcotics

The allies are struggling to combat Afghanistan's poppy crop. Some reports suggest Afghanistan supplied up to 90% of the world's opium in 2007. Poppy farmers are heavily concentrated in the south of the country and the crop is a major factor in the economic life and stability of the country. The drug trade is also a major source of funding for the insurgency as, according to some estimates, the Taliban draw an estimated 40%, or close to $100 million, of their funds annually from this industry.

The NATO/ISAF mission, from its inception, was not authorized to play a direct role in the counter-narcotics effort, such as destroying poppy fields or processing facilities. Nevertheless, NATO commanders had been instructed to provide assistance to the local counter-narcotics authorities. Britain leads the international effort to coordinate the counternarcotics assistance. The allies provide training, intelligence, and logistics to Afghan army units and police who destroy poppy fields and opium labs.[22] One former regional commander believed that the Afghan government's destruction of poppy fields was too random to be effective, and that the government has not taken decisive action to end warlord involvement in the narcotics trade. There are also reports that the government primarily destroys the crops of the poorest farmers, and leaves those of more influential families whose support is needed by the government.[23] The Bush Administration had initially urged the Karzai government to consider spraying herbicide on the poppy fields, however, the Afghan government decided against this proposal because of possible effects of herbicide on public health and the environment. No other ally reportedly supported aerial spraying largely for fear of alienating the local populations that rely on poppy cultivation for income.[24]

On September 3, 2008, the Executive Director of the United Nations Office on Drugs and Crime (UNODC), Antonio Maria Costa, briefed the North Atlantic Council (NAC) on the counter-narcotics effort in Afghanistan. Mr. Costa suggested that Afghan Army and Police efforts in the counter-narcotics strategy have largely been ineffective and that ISAF should consider expanding its support of the Afghanistan government's counter-drug effort by including the destruction of opium labs and the interdiction of drug distribution networks as part of the ISAF mission. He also suggested that ISAF focus on major drug traffickers and the drug trade along Afghanistan's borders. On October 9, 2008, at an informal meeting of NATO defense ministers in Budapest, NATO leaders agreed to authorize its ISAF forces to act with Afghan forces against opium labs and other facilities that use drugs to finance the Taliban. However, according to an article in the December 23, 2008, edition of the *New York Times*, there appears to be on-going objections by some nations that their laws do not permit their soldiers to engage in counter-narcotics operations.[25]

The repercussions of Afghanistan's poppy crop for the future of the country and for ISAF operations are extensive and complex. The Afghan government lacks the law enforcement apparatus, including a well-functioning judicial system, to combat the narcotics trade successfully. Narcotics traffickers can exploit the country's primitive transportation network, as an extensive road system is not needed to move opium to market; a small load of opium can yield a high financial return.

The opium trade has a corrosive effect on Afghan society. Former CIA Director John Negroponte told Congress in January 2007 that the drug trade contributes to endemic corruption at all levels of government and undercuts public confidence. A dangerous nexus exists between drugs and insurgents and warlords who derive funds from cultivation and trafficking. At the same time, farmers in parts of the country view the poppy as their only source of income. One component of the counter-narcotics effort is to persuade farmers to switch to alternative crops. Many crops, however, cannot compete with poppies; income from a hectare of poppies can reach $4,600 a year, while wheat, one of the suggested substitute crops, can bring only $390. Orchards might bring more money, but they require years to develop. A more extensive market infrastructure is necessary as well. Eradication of the industry without a substitute source of income would throw these farmers into destitution, and they would violently resist any effort to destroy their crops.

Another component in this effort is the status of the police and judicial systems. Some western officials in Afghanistan note that the police remain corrupt and distrusted by the population. They lack extensive training and experience, as well as transport. The police could play a key role in Afghanistan's stabilization because they, along with the Afghan army, have primary responsibility for destroying poppy fields and opium labs.[26] Police training was initially the responsibility of the Germans. The task was a daunting one, given the low pay provided by the Afghan government and the modest numbers of police used to cover a broad territory. However, difficulties in recruiting German police trainers and sub-par performance of the training program necessitated a change. Part of the problem may have been the lack of authority of the German government to order police to Afghanistan; unlike its military forces, German police must volunteer for such an assignment.[27] Some U.S. and European officials were critical of the manner in which Germany managed its task of training the Afghan police force (ANP). At the same time, former SACEUR General Jones said that while training of the Afghan army was "one of the bright stories, one of the not-so-good stories ... is the inadequacy to bring similar progress to police reform, which is the responsibility of Germany."

In May 2007, the EU accepted a request by NATO to take the lead in training Afghanistan's police. The European police (EUPOL) training mission began in June 2007 with the addition of some 120 EU police trainers who joined the 41 German trainers that remained in the program. In September 2007, the German general heading the EU police training mission reportedly quit in frustration over complications with the program, and the corruption encountered in dealing with the Karzai government.[28] In a February 2008 report by ISAF to the U.N., NATO noted that the Afghan police "still fall behind the desired level of capability."[29] In March 2008, officials at the EU suggested that the EUPOL training team could be doubled.

The EU effort has faltered thus far, for several reasons including its relations with other allied nations. Turkey has reportedly blocked any provision by NATO of intelligence to the EU and the Afghan police because (Greek) Cyprus and Malta, both in the EU, are not NATO members. Turkey is also blocking any agreement for NATO to provide protection to police who come under attack by the Taliban. Turkey's actions are a side effect of its dispute with the EU over a range of issues.

The court system remains in its infancy, with few capable jurists and attorneys.[30] The Italian government leads the effort to build a professional judicial system. In July 2007, Italy held a conference in Rome to develop a strategy to build such a system. Governments in attendance pledged $360 million to the effort over a period of several years; they linked the pledges to specific programs. Among the principles and steps that the program will seek to establish are: a code of conduct, transparency, and accountability for officials in the judicial system; and equipment, salary support, qualification requirements, and an educational system for those interested in the legal profession. A follow-up meeting was held in Kabul in October 2007 to begin implementation of these programs.[31]

Stages Three and Four: Establishing Mission and Structure

ISAF's task in Stage Three was to bring stability to the southern part of the country, where the writ of the Karzai government had been limited. In Stage Four, ISAF consolidated its responsibilities to cover all of Afghanistan. Initially, in late 2005, the allies believed that Stages Three and Four would emulate Stages One and Two by seeing a replacement of OEF forces by NATO forces in a stabilizing environment. The allies nonetheless knew that there would be several significant new challenges in both Stages. The Taliban originated in the south, in Kandahar province, and they retain their most active network there. Poppy farming is widespread in the south, particularly in Helmand province, where British troops operate, and in Uruzgan province, where Dutch troops predominate.

Stage Three came into force on July 31, 2006, after having been postponed several times due to insurgent violence and an effort to secure pledges of additional troops from allied governments. Elements of ISAF had been present in the region for several months, preparing for their mission. Stage Four began on October 5, 2006. In Stage Four, the United States transferred 10,000 to 12,000 of its own troops to ISAF, who now serve under NATO commander U.S. General David McKiernan. ISAF now has approximately 51,000 troops.

In the attempt to develop a coherent force for Stages Three and Four, the allies confronted four issues: writing a mission statement; raising troops to accomplish that mission; agreeing upon treatment of prisoners; and creating a command structure. The allies continue to address the latter three of these issues.

Mission Statement

From the fall of 2005 through early 2006, the Bush Administration argued to merge the functions and command of ISAF and OEF. Then-Defense Secretary Donald Rumsfeld asked the allies to assume counter-insurgency and anti-terror responsibilities in the southern and eastern parts of Afghanistan. Some allies balked, contending that such combat operations were OEF's task, that the U.N. resolution governing ISAF called for a stabilization operation only, and that, in some cases, the allies did not have forces available for the counter-insurgency and counter-terror tasks.[32]

In December 2005, the allies announced a mission statement for ISAF's Stages Three and Four in the form of a communiqué. They pledged to work to extend the authority of the Afghan government, primarily through development of PRTs. They also committed themselves to training the Afghan army and police, an effort in state-building meant to provide a Kabul government with reliable security forces; a formidable task because such forces were barely in existence. They further committed themselves to "supporting Afghan government counter-narcotics efforts."[33] They also agreed upon guidelines for dealing with prisoners.

The mission statement reflected European and Canadian views that Stages Three and Four operations should concentrate on reconstruction and stabilization, with initial concern over military threat at a minimum. The Taliban were relatively quiet when the allies wrote their communiqué, perhaps due to the winter weather in Afghanistan or perhaps because the Taliban were organizing and attempting to enhance their strength. In April 2006, the British Defense Secretary said that he hoped that his country's forces could deploy "without firing a shot."[34] Peter Struck, Defense Minister under the previous German government, said in September 2005 that "NATO is not equipped for counter-terrorism operations. That is not what it is supposed to do."[35] The Dutch parliament held a contentious debate in February 2006 over whether to send forces to ISAF. Some government and opposition members of Parliament opposed sending Dutch forces for combat operations; their view was clear that Dutch forces were primarily to support a stabilization mission.[36]

By spring 2006, events on the ground in Afghanistan imposed new exigencies on ISAF's mission. An attack on the Norwegian-Finnish PRT in normally tranquil Meymaneh, in western Afghanistan, in February 2006 had given an indication of an emerging problem: the need for a rapid military response capability for rescue operations. When the PRT was attacked, no NATO combat forces were in the region to protect the ISAF personnel. Other NATO forces that were nearby had caveats prohibiting their use in combat operations. Eventually a British force was found to help end the attack on the PRT. Before and after the attack on the PRT, then NATO SACEUR General Jones called upon the NATO governments to pledge forces to ISAF that would be capable of combat operations. As mentioned above, he waged a constant campaign to cajole allied governments not to place caveats on their forces that ruled out combat operations.[37]

NATO governments ultimately agreed to adjust how ISAF would fulfill Stage Three. They wrote more "robust" rules of engagement. By May 2006, then-ISAF Commander British General David Richards, described Stage Three as a "combat operation." He added that caveats affecting Stage Three and Four forces had been "reduced." He dismissed the tendency of some NATO governments to draw a line between OEF's counter-terror operations and the supposedly lowlevel counter-insurgency responsibilities that had crept into Stage Three responsibilities. He told visiting members of a NATO parliamentary delegation that counter-terror and counter-insurgency operations in Afghanistan were not always distinguishable.[38] When OEF turned southern Afghanistan over to ISAF on July 31, 2006, some OEF forces remained in the region to continue combat operations targeted against terrorist elements.

Difficulties in Raising Troops

The debate over the mission and public opinion throughout Europe continues to affect the effort to raise forces for the ISAF mission. The highest priority for any ISAF commander is to have the forces necessary along with the greatest amount of flexibility possible to provide a safe and secure environment in which the government of Afghanistan can extend its authority. Since the beginning of the ISAF mission, NATO officials have consistently experienced difficulty persuading member governments to supply adequate numbers of forces. U.S. Defense Secretary Gates has been critical of the allies at times for not providing more troops, although he has softened his tone. In December 2007 he told the House Armed Services Committee that an additional 7,500 troops were needed, in addition to the 41,700 then in ISAF. At the time, he suggested that approximately 3,500 should be trainers for the Afghan army. He also called for at least 16 more helicopters.[39] A week later, however, after a NATO Defense Ministers' meeting, he acknowledged that domestic political problems were preventing some allies from increasing their force levels in Afghanistan. Allied government officials stated privately that their populations were reluctant to follow the Bush Administration, largely due to the U.S. invasion of Iraq and subsequent criticism of the United States in Europe and the Middle East.[40] The German Marshall Fund poll noted earlier found that while 64% of those polled supported the reconstruction effort in Afghanistan, only 30% supported combat operations against the Taliban.[41]

According to NATO officials, the 2006 attack on the Norwegian-Finnish PRT awakened some governments to the continuing threat posed by instability fueled by the insurgency.[42]

Rapid-response forces eventually became available. Britain, Canada, and the Netherlands were the first to pledge forces for Stage Three. Canada was one of the first allies to recognize the need for combat forces. By a close vote in the Canadian parliament in May 2006, the government designated 2,300 troops for Afghanistan until February 2009, most of which have been sent to Kandahar province. Britain initially promised to send 3,600 troops to Helmand province by the beginning of Stage Three operations in July 2006, and has steadily increased its contribution to its current 8,000 troops. In early 2008, Germany agreed to send 200 troops to replace a Norwegian contingent in the north. In February 2008, the U.S. deployed the 24th Marine Expeditionary Unit (MEU) to southern Afghanistan.

The debate in the Dutch parliament over assigning troops to ISAF was also contentious. The Dutch population initially opposed sending forces into a combat operation. Ultimately, the Netherlands designated 1,700 troops for duty in ISAF's Stage Three and Stage Four operations.

Despite these pledges, the upturn in violence in 2007 and 2008 led U.S. and NATO commanders in Afghanistan to conclude that they needed about three more brigades (20,000 troops) to be able to stabilize the still restive southern sector.

At the April 2008 NATO Summit at Bucharest a key objective of several allies with combat contingents in Afghanistan was to persuade other governments to send more forces. When the allies issued their "strategic vision" statement on Afghanistan, the allies agreed to a shared longterm commitment, something that some allies had theretofore resisted stating publicly, but they did not promise to contribute an equitable share of combat forces. Part of this inequity is attributable to NATO's own budget rules. When a member state agrees to deploy troops to a NATO operation, that nation must pay the costs associated with that deployment. Thus, there is a built in disincentive for nations to agree to commit any troops to a mission or to increase the number of troops already deployed. This problem complicates attempts by leaders of fragile governments or coalition governments to convince their legislatures and publics to support a deployment and the costs associated with that commitment.

In 2008, the Canadian government threatened to withdraw its forces in 2009 if a commitment of at least 1,000 new combat troops was not made by the allies. President Bush pledged to increase U.S. forces in Afghanistan by 5,000 additional troops by the end of 2008. France agreed to send 720 combat troops. Germany agreed to deploy an additional 1,000 troops to the northern sector pending approval by the German Parliament in October when the current German mandate was to have expired. Poland, the Czech Republic, and several other allies pledged smaller contingents, allaying Canadian concerns to some degree. However, allies with forces in harm's way continued to criticize other allies that will not send combat forces or commit them to areas where the Taliban are more active.

As the perception of deterioration in Afghanistan continued, it was reported in September 2008 that both the U.S. military and NATO were conducting a number of different strategy reviews. Among the issues under review was how to prevent the movement of militants across the Afghanistan-Pakistan border. Although some of the reviews have been completed, it is unclear if or when any major changes in policy would be initiated. U.S. officials at the time said more U.S. and partner forces were needed, and U.S. officials were trying to identify more forces to go to Afghanistan. In September 2008 President Bush announced that at least 5,000 more U.S. forces would be sent to Afghanistan by early 2009. However, ISAF Commander, General McKiernan recently suggested that the effort still needed about 20,000 additional troops. The incoming Obama administration has suggested that an additional 20,000 - 30,000 U.S. troops will ultimately be deployed to Afghanistan.

The new U.S. Administration has already hinted that it will ask other partner countries to contribute additional forces to the ISAF mission. However, in a sign of how stretched some allies are or how reluctant others will be, U.K. Prime Minister, Gordon Brown on December 16 indicated that he could only contribute an additional 300 British troops in 2009. In November, a U.S. House congressional delegation visiting Italy was told that the Italian troop commitment to Afghanistan could not be increased further. Also in November, the Spanish Foreign Minister told a meeting of the NATO Parliamentary Assembly in Valencia that Spain would not send additional troops to Afghanistan.

The reluctance of the NATO allies to commit additional troops to the ISAF mission is being driven in part by the opposition of many of Europe's citizens who see little progress in Afghanistan and in part by the realities of the global economic crisis that is currently having a negative impact on national budgets. This could complicate attempts by the new U.S. Administration to create the kinds of conditions in Afghanistan that could lead to a greater stabilization of the country. The NATO allies are expecting President Obama to ask for more allied troops and are preparing for that discussion at the April 4 NATO summit in Germany and France. These additional troops would be used to train the Afghan security forces, to try to stabilize the still restive southern sector, and reverse the deterioration of the eastern sector and the areas around Kabul. Some equate this to the Afghanistan equivalent of the U.S. "troop surge" that has been credited with greatly reducing violence in Iraq.

For some, the concern over the troop issue means that the extra forces will likely be contributed almost entirely by the U.S. This could further add to the "Americanization" of the conflict in Afghanistan and could provide less of an incentive for NATO allies to send additional troops. This development could result in an increase in negative criticism of NATO and Europe within the United States. Some critics of the Alliance counter with the argument that if every one of the NATO member countries, with the exception of the United States contributed just 200 additional troops, the ISAF Commander would have 5,000 new assets to deploy. Similarly, just 400 troops per nation would provide ISAF with 10,000 troops. Even eliminating the request for additional troops from those nation already deploying thousands of troops, a commitment of somewhere between 200 and 400 troops by the rest of the Alliance would provide a sufficient new force. These critics point to countries such as Bulgaria, Romania, the Czech Republic and Hungary that have large armed forces but contribute less than 500 troops each. Similarly, critics ask why a nation such as Turkey that deploys 20,000 troops in northern Cyprus can only provide 800 troops as their commitment to the Alliance mission.

Disagreements over Treatment of Prisoners

There was a contentious debate among the allies over the December 2005 final communiqué guiding NATO operations in Afghanistan. Most of the allies were critical of U.S. abuse of prisoners at the Abu Ghraib prison in Iraq; they extended this criticism to the U.S. detention policy at Guantanamo Bay, where some prisoners captured in Afghanistan had been sent since 2001. These allies contended that the Bush Administration was ignoring the

Geneva Convention governing treatment of prisoners taken in combat, and that the issue was a significant one among their publics and in their domestic political debates.[43]

These allies insisted that the communiqué explicitly address the issue of treatment of prisoners. The final document contained the statement: "In addition to NATO's agreed detention policy for ISAF, which is and remains consistent with international law, we welcome initiatives by Allies to assist the Afghan authorities in the implementation of international standards for the detention of prisoners."[44]

The allies also agreed that prisoners taken by ISAF should be turned over to the Afghan government. Some allied governments reportedly told the Afghan government that they did not wish such prisoners to then be transferred to the United States. The Afghan government reportedly insisted upon its sovereign right to determine the disposition of prisoners in its custody. A new problem has arisen over allegations that Afghan officials have tortured detainees turned over to them by ISAF forces.[45]

Command Structure: Coordinating ISAF and OEF Operations

NATO's discussion over the command structure for Stages Three and Four in Afghanistan had reflected the U.S. desire to see the allies more fully embrace counter-insurgency tasks. Reluctance on the part of some European governments to clash with the Taliban and warlords was evident during these discussions.

Although the allies agree on ISAF's mission, they differ on how to accomplish it. From at least 2004, the former Bush Administration had consistently urged the allies to assume more responsibilities in the fight against insurgents and terrorists in Afghanistan. By late 2005 the Administration was urging that ISAF and OEF be merged under one command. Britain, Germany, and France were the principal allies opposing the U.S. idea to merge the commands. They did so for differing reasons. Britain and Germany wished to preserve ISAF as a stabilization, and not combat, mission and because German forces in ISAF were trained only for stabilization, and not for counter-insurgency operations.. Britain, leading the ISAF anti-narcotics effort, wished to ensure that the initiative remained in the political sphere.

The French view was somewhat different. The French government was close to the Administration view that some combat operations against the Taliban and other elements would be necessary. At the same time, France was concerned that the Administration, after having a U.S. commander in place to guide all military activity in Afghanistan, might use NATO as a "toolbox" to accomplish Washington's broader global objectives. Specifically, Paris was concerned that the Administration would designate more U.S. units from Afghanistan to be sent to Iraq, and leave the allies to stabilize Afghanistan. Administration officials insisted publicly and privately that they had no intention of sharply reducing forces in Afghanistan, and in fact had increased U.S. forces there.[46] Nevertheless, the government of President Nicolas Sarkozy, as noted earlier, decided to increase its combat contingent in Afghanistan.

In attempting to resolve the issue of command structure, the allies sought to address practical problems for the two operations. ISAF and OEF operate in contiguous areas, but there has been no clear dividing line between regions where the Taliban and al Qaeda are active, and the relatively stable regions of the country. A weakness of ISAF had been deficient capability for rapid response rescue, should soldiers and civilian personnel find themselves under fire. In September 2008 at a hearing before the Senate Armed Services Committee, U.S. Defense Secretary Robert Gates, testified that he believed there were still questions regarding the efficacy of having two lines of command.[47] He noted that he was considering a plan to place almost all U.S. troops, including those performing OEF anti-insurgent missions, under General McKiernan's NATO/ISAF command,

Today, a "synergy" exists between the two operations which allows each operation to support the other in times of emergency. U.S. General David McKiernan, now commands both the OEF and ISAF operations which has served tocreate unity of command, and to improve flexibility of deployment of U.S. forces throughout the battlefield. General McKiernan and his successors report to U.S. Central Command (CENTCOM, now headed by General David Petraeus, formerly top U.S. commander in Iraq) not only to NATO headquarters. However, U.S. officials say that the OEF and NATO/ISAF missions will not formally merge, meaning that there will still be separate U.S. operations against high value targets and other militant concentrations.[48]

Allied Viewpoints

Once the allies reached consensus on ISAF's mission for Stages Three and Four, they began to differ on how to accomplish it. The previous section analyzed allied views in establishing and implementing the mission and structure of Stages Three and Four. This section discusses the developing views of allies as Stage Three and then Stage Four moved forward. Allied views began to change between the time of the December 2005 NATO communiqué describing ISAF's mission and July 2006, largely due to the surge in Taliban activity. For purposes of analysis, the range of views begins with governments most hesitant about the use of combat forces in Afghanistan and proceeds with governments that believe that a more forceful military hand will be necessary to stabilize and rebuild the country.

Germany: Reconstruction is the Priority

After coming to power in October 2005, Chancellor Angela Merkel's coalition government initially expressed a more decisive commitment to securing stability in Afghanistan than its predecessor. Berlin has long advocated a shift in its and NATO's Afghanistan strategy toward civilian reconstruction and development projects, army and police training activities, and enhanced political engagement with Afghanistan's neighbors. Under the preceding Schroeder government, Berlin was adamant that German forces would not engage in combat operations; according to NATO officials, the German caveat against combat had limited the alliance's ability to integrate German forces with those of other allied governments. Under the Merkel government, German forces are authorized to engaged in combat if in defense of German positions but they are still prohibited from engaging in counter-insurgency operations.

Germany now has approximately 3,500 forces in ISAF trained for stability operations in the northern part of the country where they lead two PRTs, one in Kunduz and one in Feyzabad. Some officials from other allied governments and the EU have criticized the existing restrictions on German forces and the capabilities of those forces. These officials say that German troops and civilians rarely venture beyond the perimeter of their PRTs due to concern that they might arouse Afghan public criticism or come into contact with armed elements. German troops reportedly do not go on extended patrols and do not respond to local security incidents. Critics of the German approach say that it is important to engage local officials and demonstrate that NATO has an active approach to rebuilding the country and persuading the Afghan population that the alliance is serving a constructive role.[49] However, even this area has become more dangerous as the Taliban increase operations throughout the country. For instance, on October 20, 2008, a suicide bomber in Kunduz killed several civilians along with two German soldiers. German forces are authorized to engage in combat operations as part of their defense of the northern sector but they are not deployed to conduct counter-insurgency operations.

At NATO's Riga summit Germany agreed to send troops to assist allied forces in an emergency. In spring 2007, the German government assigned six Tornado aircraft to Afghanistan for use in surveillance operations. In October 2007 when the *Bundestag* renewed the commitment to keep German forces and Tornado aircraft in Afghanistan for another year, Chancellor Merkel rejected an appeal by the NATO Secretary General to send some of Germany's forces to the south for stabilization operations. Since then, however, Germany has agreed to send troops to other parts of the country to assist allied forces in emergency situations.

As noted above, in June 2008, Berlin announced that it would seek approval to increase troop levels in Afghanistan by up to 1,000. On October 7, 2008, the German government extended the German troop commitment to Afghanistan and agreed to send the additional 1,000 troops to Afghanistan. On October 16, 2008, the German *Bundestag* approved the government's decision in what was considered a fairly non-controversial debate. In approving the additional deployment of German forces, however, the *Bundestag* made it clear that no additional troops beyond the additional 1,000 would be approved, and that no special forces troops would be assigned to the OEF counter-insurgency operation. The additional troops are expected to boost Germany's efforts in northern Afghanistan, with a stated aim of tripling the amount of training Germany gives to Afghan troops.[50] In December, Berlin announced that it would provide 3 million euros to aid the Afghanistan police force. The additional funds will be provided through the United Nationsbacked Law and Order Trust Fund.[51]

While Chancellor Merkel and her Foreign Minister Frank-Walter Steinmeier (also Merkel's opponent in national elections scheduled for October 2009) have consistently expressed their support for the ongoing German military engagement in Afghanistan, public support in Germany for the Afghan mission has steadily declined. In 2002, 51% of those polled supported German involvement in Afghanistan's stabilization; as of October 2007, that figure had declined to 34%. In September 2008, a new survey of public opinion conducted by the German Marshall Fund found that while German support for the Afghanistan mission continued to be lukewarm, support among the population for combat operations against the Taliban has declined to around 36%.[52] Low public support for the ISAF mission and some political opposition from within Chancellor Merkel's coalition have dampened expectations. According to some observers, the German population has serious doubts about Germany's role in Afghanistan and they are beginning to feel Germany does not have a winnable strategy for Afghanistan. Some observers also fault Chancellor Merkel for failing to lay out the importance of the Afghan mission to the German people.[53]

The Netherlands: An Increasingly Decisive Position

Dutch forces numbering approximately 1,700 are concentrated in the south, in Uruzgan province, one of Afghanistan's most unstable regions and an area that has seen considerable Taliban activity since the spring of 2006. The debate in the Dutch parliament over assigning troops to ISAF was contentious. The Abu Ghraib prison scandal and U.S. treatment of prisoners at Guantanamo were important issues in the Dutch debate over Afghanistan. Dutch officials say that "the rules of the road in fighting terrorism" are not clearly agreed upon in the alliance. For this reason, Dutch officials were initially reluctant to have their forces closely associated with U.S. forces in Afghanistan. The Netherlands was the principal proponent of the section of the December 2005 NATO communiqué detailing allied treatment of prisoners in Afghanistan.[54]

The Dutch population initially opposed sending forces into a combat operation. Ultimately, the Netherlands deployed significant troop levels for duty in ISAF's operations. Dutch troops have grown increasingly engaged in providing security, in tandem with an active and well-funded reconstruction effort. In December 2007, the Dutch parliament agreed to keep troops in Afghanistan, but to begin a withdrawal in August 2010 until all Dutch forces are withdrawn by December 1, 2010. The parliament expressed dismay that more allies have not been forthcoming in providing forces for southern Afghanistan.

In the Dutch view, ISAF's purpose is "to provide a secure and stable environment for reconstruction." The government's policy has been that measures of "defense, diplomacy, and development" are key to ISAF's success. When necessary, Dutch troops will use force to subdue the Taliban to build stability so that reconstruction projects may take hold. A growing number of combat engagements, occasionally along with U.S. troops, has occurred since late summer 2006, and Dutch forces have suffered casualties.[55] The Netherlands endorsed the "synergy" of ISAF and OEF commands and has made available four F-16s for missions in both ISAF and OEF. The aircraft may be used for missions from intelligence gathering to close air support.

Dutch officials offer a strategic approach to Afghanistan's problems. They believe that the alliance must make a more concerted effort to engage regional countries—above all, Pakistan, India, and Iran—to bring stability to the country, a view given increasing attention in some allied capitals after the assassination of Benazir Bhutto. These officials are concerned that NATO's military operations, including air strikes, are alienating the Afghan population. They have advocated the creation of a general fund to rapidly compensate local victims of mistaken attacks by NATO forces. In addition, they advocate a common approach in NATO and the EU to the problems presented by the drug trade. In the Dutch view—echoed by Italy—NATO must emphasize reconstruction more than combat operations.[56]

Others counter this argument by saying that "there can be no reconstruction without security." The Taliban must be cleared out before reconstruction can proceed. Many in the ISAF command share the Dutch view that NATO should build roads and other economic infrastructure to help create an economy to give Afghans promise of a future.[57]

As stated previously, the Dutch give their funding for PRT reconstruction activities directly to the Afghan central government, mainly through U.N. and World Bank channels. Dutch officials note the contrast with the U.S. approach, which is to bring in a "turnkey" operation in which U.S. officials are trained to undertake reconstruction projects, using U.S. manpower and equipment. The Dutch argue that the Karzai government itself must undertake responsibility for planning and implementation of projects to rebuild the country. Only in this way, the Dutch believe, can the Afghans learn good governance and management of their own affairs. The Dutch are directly involved in some projects, providing clean water to villages and almond trees and seeds to farmers for alternative crops, for example. Some U.S. officials believe that the Dutch practice of providing assistance funds directly to the Afghan government has led to the money being spent on other governmental purposes or landing in the pockets of corrupt Afghan officials.[58]

The United States, Britain, and Canada: A Broad Mandate

The governments of the United States, Britain, and Canada share similar views on how ISAF should fulfil its mission. They have sent combat forces to Afghanistan, maintain PRTs in the most unstable parts of the country, and have engaged the Taliban resurgence aggressively. Many of the British and Canadian forces for Stage Three began to arrive in Afghanistan in spring 2006, and worked under OEF command fighting the Taliban. On July 31, 2006, most of these forces were "rebadged" as NATO forces serving ISAF's Stage Three mission.

The United States now has approximately 19,000 troops in ISAF. There has been an ongoing debate in the Pentagon over whether a possible draw-down of U.S. forces in Iraq could open the door for more U.S. forces in Afghanistan. Former President Bush agreed to increase U.S. forces in Afghanistan by 5,000 by the end of 2008, necessitated at least in part by some allies' refusal to agree to a greater sharing of the burden.[59] An additional 20,000 - 30,000 U.S. military forces could be sent to Afghanistan during 2009.

U.S. officials believe that ISAF must undertake tasks "from the lowest level of peacekeeping to combat operations against the Taliban and warlords." OEF's task should be counter-terrorism against al Qaeda. These officials concede that the line between the two operations is blurred, given that OEF has been fighting both an insurgency led by the Taliban and searching for al Qaeda.[60] Some allied governments believe that the U.S. combat effort is overly aggressive and, in some instances, has been counterproductive. President Karzai has said that air strikes have sometimes been poorly targeted and have carelessly killed civilians, which he believes may be alienating the population in some areas of the country. In July 2007, NATO announced a new policy. ISAF would postpone a combat response, where possible, when civilians are present near the Taliban; in addition, ISAF aircraft will use smaller bombs to limit damage to an area.[61] In October 2008, NATO/ISAF further refined that policy by suggesting that NATO forces would disengage when the need for air strikes could endanger local civilian populations.

The British view largely mirrors the U.S. view of NATO's role in Afghanistan. From a cautious position on ISAF's mission in early 2006, the British government has adopted a more aggressive stance, as a result of the increase in Taliban activity in southern Afghanistan. Britain has ISAF and OEF contingents, and its combat aircraft support both missions. Most of Britain's ISAF troops, numbering approximately 8,700 are located throughout the country with approximately 4,200 in the south. British forces in the south are largely in Helmand province, the principal poppy-growing region in the country. British forces have an "inkblot" strategy, in which they clear an area of Taliban, then undertake reconstruction projects, such as road building, moving out from a village into the countryside.[62] Britain has a clearly vested interest in ISAF's stabilization mission, not only out of concern that terrorist activity has emanated from south Asia but because most of the heroin found in the United Kingdom comes from Afghanistan. U.S. officials believe that Britain's PRT in Helmand province is well-funded and concentrates on local governance and economic development.[63]

At the same time, a debate over the proper balance between combat missions and reconstruction continues in Britain. Prime Minister Brown's government now reportedly believes that more emphasis must be given to an effort to reconcile elements of the insurgency with the Karzai government. The British government reportedly believes that there are "hard-core" Taliban elements incapable of accepting the Karzai government, but that there are other levels of Taliban, not affiliated with Islamic extremism, that can be persuaded to lay down their arms. A key component of such an approach would be a successful reconstruction effort that would provide jobs and broadened economic growth.[64] Britain had reportedly hoped to increase its force contingent at the Bucharest summit, but increased violence in Iraq has kept combat elements tied down there.

After the Bucharest summit in April 2008, the Brown government came under increasing political and economic strain. The opposition has called for general elections earlier than those scheduled for 2010. Some in Brown's own government had suggested he step down as leader of the Labor Party. Brown has recovered somewhat recently as a result of his actions in response to the global economic crisis but his recent decision to only send an additional 300 troops to Afghanistan continues to reflect the opposition to the war in the U.K. The decision by Canada and the Netherlands to begin withdrawing troops from Afghanistan by 2010 could become a campaign issue for Brown in the run up to the 2010 elections.

Canadian forces joined U.S. and British forces in the summer and fall 2006 OEF combat operations against the Taliban in southern Afghanistan. Some of these operations, led by Canadian teams, were joined by Afghan army (ANA) elements in Kandahar province. The Canadians eventually wish to turn over such operations to the ANA. Some of the Canadian forces assigned to OEF were transferred to ISAF's Stage Three operations on July 31, 2006, and Kandahar province is their principal region of responsibility. Canada leads a PRT in the province.

There has been a vigorous debate in Canada over the country's involvement in Afghanistan. In 2006, by a narrow vote of 149-145, the Canadian parliament approved Ottawa's plan to commit 2,300 troops to ISAF until February 2009. Public support for the mission has fallen, however. In 2002, 66% of those polled supported sending Canadian forces to Afghanistan, but only 44% supported the two-year extension until 2009. By April 2007, support for keeping Canadian forces in Afghanistan had dropped to 52%. While Canadians appear to support their country's long involvement in U.N. peace operations, the need for combat operations in Afghanistan has eroded support for the ISAF mission. When the alliance pledged more combat forces for southern and eastern Afghanistan at the Bucharest summit, Ottawa withdrew its threat to remove its troops in 2009. On March 13, 2008, the Canadian parliament extended the commitment to keep troops in Afghanistan until 2011.[65]

France: Combat and Stabilization

The French government believes that ISAF must be a combat force that buttresses the efforts of the Afghan government to build legitimacy and governance. Unlike German forces, for example, many French forces are trained both for combat and stabilization. As of September 2008 France has 2,700 troops in ISAF; most are in a stabilization mission in Kabul and in army training missions elsewhere in the country. French officials express concern that ISAF will fail "if not accompanied by increased capacity by the Afghan police and judicial system."[66]

French President Nicolas Sarkozy has reaffirmed Paris's commitment to ISAF. In 2008 France moved 6 Mirage fighter bombers from a French base in Tajikistan to the NATO base in Kandahar, in southern Afghanistan. These jets are used in intelligence and close air support missions; their relocation to Kandahar will allow them to spend more time in the air on mission rather than on the long return to Tajikistan for resupply.[67] France also supplies C135 tankers to refuel French and other allied aircraft. France has built 4 operational "OMLTs," a term used to describe a joint allied and Afghan combat force, and participates in another with Dutch forces. These forces are in the east and south where combat is at the highest levels. U.S. and French forces are jointly training Afghan special forces teams.

As noted above, President Sarkozy pledged an additional 720 combat troops for Afghanistan at the NATO Bucharest summit in April 2008. Two hundred are special forces, and some of these may join U.S. forces in OEF; the rest are in mobile combat units. These troops will be under U.S. command in eastern Afghanistan where, according to NATO, Taliban operations "continue at a high level."[68]

The opposition Socialist Party in France has strongly criticized Sarkozy's decision to increase French force levels in Afghanistan. The Socialist leader in the National Assembly characterized the decision as asking "France to support in Afghanistan the American war burden in Iraq" as part of the French president's "Atlantic obsession;" in this view, European forces in Afghanistan free the United States to send or keep forces in Iraq, a war that is highly unpopular among the French public. Prime Minister François Fillon responded that in fact the troops will be sent to Afghanistan as part of a NATO "common strategy." President Sarkozy has described ISAF's mission as one to counter global terrorism.[69] Criticism of Sarkozy's commitment increased in August 2008 after a French combat patrol was ambushed by Taliban forces resulting in 10 French casualties. President Sarkozy visited French forces soon after the battle and reiterated France's continued commitment to the ISAF mission.

On September 22, 2008, the French Parliament, at the insistence of the Socialists, debated the continued presence of French military forces in Afghanistan. By large margins, both the National Assembly and the Senate voted to continue to support French participation in Afghanistan. After the vote, President Sarkozy announced that an additional 100 troops would be sent to Afghanistan along with helicopters and ariel drones.

The French government, mindful of civilian casualties and Afghan criticism of ISAF, is emphasizing more restrictive rules of engagement for its forces. Its troops have been instructed to use force "proportional" to a threat, to avoid bombing civilian infrastructure, and to have "visual recognition" of a target before releasing bombs.[70]

The Afghan mission has marked important changes in French NATO policy. France supported the invocation of Article V, NATO's mutual security clause, after the attacks of September 11, 2001, on the United States. Those attacks were decisive in the French government's change of position on NATO's "out-of-area" responsibilities. For many years, Paris had argued that NATO was a European security organization, and must only operate in and near Europe. After September 11, the French government embraced the emerging view that NATO must be a global security organization able to combat terrorism and WMD proliferation around the planet. French officials say that ISAF is NATO's most important mission.[71]

Since the late 1990s, NATO has urged member governments to construct more "deployable," expeditionary forces, and gave the notion a concrete base in the Prague Capabilities Commitment (PCC) in 2002, when allies pledged to develop capabilities such as strategic lift, aerial refueling, and more special forces.[72] Among the European allies, France has made considerable progress along this path. As noted above, French aerial tankers refuel not only French aircraft in the Afghan theater, but U.S., Dutch, and Belgian aircraft as well. These capabilities contribute to the improving integration of NATO forces in the Afghan theater, according to U.S. officials, and to the ability of ISAF and OEF to share capabilities and command.[73] U.S. officials give French forces high marks for their ability and their willingness to fight.

The French government has clearly defined its interests in Afghanistan. French officials argue that the allies must commit to a long effort to assist the Afghan government in eradicating the opium industry, in part because heroin finds its way into western societies, in part because it fuels terrorist groups. Ultimately, French officials believe that the Afghan government itself must learn to govern the country, and that NATO and partner states cannot do this for Kabul. To this end, the French have a contingent in place that assists in training the Afghan army. France does not believe that PRTs can play a meaningful role in Afghanistan, and believes that the Karzai government must itself exercise the initiative and build good governance to gain the confidence of its people. France does not accept the view, held by some U.S. officials but nowhere present in NATO's ISAF mission statement, that part of NATO's brief is to build democracy in Afghanistan. In the French view, Afghanistan is a highly diverse ethnic state with no tradition of democracy; at best, for the foreseeable future, a more representative and tolerant society can be built.[74]

France also contends that the EU and other civilian institutions, such as the U.N. and the World Bank, are more suited to undertake development projects than NATO. In Paris' view, NATO should concentrate on collective defense.

Congressional Action

A bipartisan consensus continued to support the Afghanistan mission during the 110th Congress. The Afghan Freedom Support Act of 2002 (P.L. 107-327), as amended, authorized U.S. aid for reconstruction, military operations, counter-narcotics efforts, election reform, and human rights assistance. A succession of appropriations bills has met or exceeded authorization targets.[75] Since the 9/11 attack, Congress has appropriated over $176 billion for Afghanistan. In hearings during the first session of the 110th Congress, Administration officials told Members that the United States spends approximately $2 billion a month in Afghanistan on troops and reconstruction.

On June 19, 2008, the House passed the supplemental appropriations bill and on June 26, the Senate passed its version. On June 30 the President signed the FY2008-FY2009 Supplemental Appropriations bill into law (P.L. 110-252). The law provided $31 billion for the conflict in Afghanistan. The bill also provided $1.3 billion for economic reconstruction in the form of Economic Support Funds (ESF), primarily to strengthen regional governance, health care and education, development of the rural economy, and civilian infrastructure.[76] An additional $35 million was provided to support the counter-narcotics programs in Afghanistan.

On July 15, 2008, former-Senator Biden and Senator Lugar, the Chairman and Ranking Member of the Senate Foreign Relations Committee, introduced S. 3263, a bill that would significantly increase non-military aid to Pakistan. The assistance would, among other things, be used to improve Pakistani counter-terrorism capabilities and ensure more effective efforts are made against Taliban and al Qaeda forces using Pakistan as a springboard for launching military and terrorist attacks into Afghanistan.

Several hearings were held during the 110th Congress that addressed a range of Afghanistanrelated issues, including troop levels, command and control arrangements, counter-narcotics efforts, PRTs, and others. During the 110th Congress, congressional committees continued to press Secretary Gates and other officials to provide Congress with a more detailed accounting of ISAF's operations, and urged the Administration to persuade the allies to provide a greater proportion of ISAF's forces. In addition, the National Defense Authorization Act for Fiscal Year 2008 (P.L. 110-181) established three new reporting requirement: a twice yearly report from the President on progress toward security and stability in Afghanistan; an annual report from the Secretary of Defense on a long-term, detailed plan for sustaining the Afghan National Security Forces; and a one-time requirement for a report from the Secretary of Defense on enhancing security and stability along the Afghanistan-Pakistan border.[77] The 111th Congress is likely to continue to pursue oversight of the operation in Afghanistan. Hearings on the overall status of the conflict, the NATO commitment, the ability of the Kabul government to establish nation-wide authority, and the problems associated with the unsettled Pakistan-Afghanistan border are likely topics for those hearings.

ASSESSMENT

Afghanistan's long history without an accountable central government able to extend its reach over the country's difficult geographic and political terrain continues to present the allies with problems rivaling the threat of the Taliban. For some, Afghanistan's political transition was completed with the convening of a parliament in December 2005. However, after seven years neither the government in Kabul nor the international community has made much more than incremental progress towards its goals of peace, security, and development. According to a March 2008 report issued by the Atlantic Council of the United States, the situation on the ground has gradually settled into a strategic stalemate. NATO and Afghan forces cannot eliminate the Taliban threat by military means as long as they have sanctuary in Pakistan, and the civil development efforts are not bringing sufficient results.[78] With this reality, there have been increasing calls for the Karzai government and the US/NATO leadership to consider reaching out to moderate Taliban forces and sympathizers inside Afghanistan to explore the idea of a cease fire and coalition government. Meetings between the Kabul government and some elements of the Taliban were held during the summer of 2008 but it would appear at this point that the Taliban is too disjointed of a movement to provide any realistic political settlement.

The declining fortunes of the Karzai government also present a difficult obstacle. NATO is attempting both to respect the policies of a nascent representative government and to urge it forward to better governance. The Karzai government's own problems are apparent: discontented warlords, endemic corruption, a vigorous drug trade, the Taliban, and a rudimentary economy and infrastructure. In the view of former NATO Deputy Commander, General Eikenberry, "The enemy we face is not particularly strong, but the institutions of the Afghan state remain relatively weak."[79] There is a widespread view that President Karzai is losing the confidence of the Afghan people; he blames the slow pace of reconstruction and insufficient financial support from the international community. General Ed Butler, the former commander of British forces in Afghanistan, said back in May 2006: "This year we need to be seen to be making a difference. It is a real danger that if people do not feel safer, we may lose their consent." In his view, poor governance and not the Taliban insurgency was the country's central problem, a view widely reflected by other officials from NATO governments.[80] NATO, in this view, must prepare to deal with successive governments of unknown composition and policies should the Karzai government fail to endure. New Presidential elections will be held in 2009 and several candidates have already declared their intention to oppose President Karzai's re-election. U.S. and NATO political and military leaders must be prepared to walk a fine line during the election campaign as they continue to prosecute the war effort and work with the existing government.

NATO's effort to assist the Karzai government in weakening the narcotics trade demonstrates a central dilemma of ISAF's mission. The allies must fight an insurgency tied to the opium industry with forceful means while at the same time attempt to win the confidence of the Afghan people through reconstruction of the country and by providing poppy farmers alternative ways to make a living. In this view, "breaking down suspected insurgents' doors in the morning [makes] it difficult to build bridges in the afternoon."[81] While NATO officials state publicly that allied forces have not been burning poppy fields, depending instead on the Afghan army and police to do the job, farmers are well aware that it is ISAF that supplies the intelligence, training, and logistics enabling government security forces to attack the industry, the lifeline of many poor Afghans.[82]

NATO's training of Afghan officials has made measured progress in some areas, and very little in others. Although the Karzai government has complained that NATO is not building a sufficiently large army fast enough, most allies believe that substantial progress has been made in developing a professional and reliable force that will eventually equal approximately 125,000 troops. Since the beginning of Stage Three, British, French, and Canadian troops have reportedly given more and more responsibility to the ANA in joint operations.[83] As of September 2008, the ANA had 70,000 troops.

NATO and the broader international community are now making a more substantial effort to reform the judicial system and build an effective police force. Italy has successfully urged donor nations to provide more funding to build a judicial system and to begin implementation of specific programs using the funds. The EU has assumed responsibility for training the police, and put professional trainers on the ground in June 2007, an effort yet to bear significant fruit.

NATO faces complex issues in its own ranks and on the ground in Afghanistan that are likely to concern ISAF over the next several years. Although the allies agree on their overall mission to stabilize the country, they often differ on the means to reach that objective and on the amount of resources to be made available. As a result, NATO commanders continue to have difficulty persuading allies to contribute forces to ISAF or to provide NATO forces the appropriate equipment for their tasks. Despite past pleas for more troops, and an anticipated new effort by the U.S. administration to secure commitments for additional troops, it does not seem likely that significant troop increases will come from NATO member countries. The Obama Administration is likely to continue to work behind the scenes to have NATO leaders announce at least a symbolic contribution of additional troops when the Alliance meets in April for its 60th anniversary summit.

The issue of military equipment continues to remain an unresolved problem. Many allied forces lack night-vision equipment, or the technology necessary to detect roadside bombs. The current global economic crisis has made it difficult for some militaries to even consider buying the requisite equipment. Some NATO governments send forces inappropriate for the task, forces that are heavy on support functions but light on combat capability. These governments tend to be reluctant to send their forces into the field to confront the Taliban and to control warlords and their militias. The result, in this view, has been that British, Canadian, Dutch, Danish, French and U.S. forces bear a disproportionate share of the most dangerous tasks.[84] For some allies, it is clear that fighting the Taliban, warlords, and the narcotics trade can prevent the return of the Taliban, al Qaeda, other or radical Islamic groups inimical to western interests. For others, the sooner the Afghan government and the civil sector can win the hearts and minds of the general population through economic development and the efficient provision of services, the faster stability will take hold.

The allies had long reached a consensus that reconstruction is the key to building a viable, functioning Afghan state. Officials in allied governments repeatedly point to the need for more road building to extend the reach of Kabul and to provide the infrastructure to diversify and strengthen the economy of a country lacking the capacity to develop enduring market practices. U.S. General Eikenberry, former Deputy Commander in Afghanistan, when asked by a congressional committee what he needed to build a stable society, responded, “Would I prefer to have another infantry battalion on the ground of 600 U.S. soldiers or would I prefer to have $50 million for roads, I’d say ... $50 million for roads.”[85] His view has been echoed by calls from the NATO Secretary General for allies and international institutions to provide more funds for reconstruction.

Political differences within the alliance over how to manage Afghanistan’s future are apparent in ISAF’s operations. The allies’ description of PRTs as the “leading edge” of their stabilization effort masks a divergent reality. Some PRTs are clearly effective, building needed infrastructure and by most accounts gaining the confidence of local populations. Others, in the view of some U.S. and European officials, are no more than showcases, aimed more at demonstrating an ally’s desire to participate in an important NATO mission than at producing concrete results for the stabilization plan.

The United States has made an evident effort through its PRTs to engage local Afghan leaders and the general population to convince them of the worth of ISAF’s mission. While some progress has clearly been made, several U.S. officials have noted that Afghanistan is a society where personal contact and developed relationships are critical in building trust and in persuading Afghans to pursue better governance. The short rotations of some allied forces impede this effort. Some allied governments, however, are now sending troops into Afghanistan for two-year rotations, which provide a better opportunity to gain the confidence of the population.

Some observers have indicated that while some governments have pledged money in the past many have not yet contributed to the reconstruction effort. At an international conference to support Afghanistan convened in Paris in June, 2008, delegates pledged $20 billion to support the Afghanistan National Development Strategy. Key allied governments say that they are committed to staying for a period of years to stabilize the country. Some EU officials believe that five years or more will be necessary to build a market economy and proficient governance.[86] Some allies that were reluctant to commit additional troops to the ISAF mission had suggested that they may be willing to contribute additional financial or technical support through the EU effort in Afghanistan or the UN. However, the recent global economic crisis has forced most of those nations to pull back from making significant financial contributions.

Cohesiveness of command is another lingering issue. While the allies reached agreement on a command structure linking ISAF and OEF, some observers believe that national commands will preserve the authority to make final decisions about use of their forces. The Dutch parliamentary debate clearly signaled this inclination.

An additional problem for many of the allies is a growing feeling of an "out-of-area" fatigue in Afghanistan and a lack of public support for continued involvement in the ISAF mission. Many of the allies have a difficult enough problem trying to address the issues of troop commitments and national caveats for their troops already in Afghanistan. Now, however, many are facing a growing restlessness among the public who see little accomplishment and who are calling for bringing the seven year war to an end or at least withdrawing troops from the theater. Many Europeans no longer see Afghanistan as a threat to Europe and in the wake of the Russian conflict with Georgia in the summer of 2008, believe NATO needs to refocus its responsibilities.

For their part, the allies desire to stabilize the country and to prevent the return of a terrorist state has led to an ongoing general consensus that if Afghanistan cannot be stabilized and made more secure, the future credibility and relevance of NATO will come into question. Press reports state that the allies produced a classified document at the Bucharest summit that lays out a five-year plan for ISAF to stabilize the country and turn most combat operations over to the Afghan National Army.[87] Subsequently, in September 2008, the new Commander of the U.S. Central Command, General David Patreus, indicated that he was conducting a total review of the Afghanistan war strategy with the intention of developing a more comprehensive U.S. approach to dealing with that conflict along with ideas for the ISAF mission. If ultimately successful, ISAF can help to build a state that is relatively stable, no longer a source of international terrorism, and one that attempts to diminish a narcotics trade that is a threat to European societies.

ISAF may be having a residual, positive effect on the militaries of some NATO members, particularly new member states. U.S. military personnel say that true reform of new members' militaries can best take place in the field, under difficult conditions, and through operations with more experienced NATO militaries. By several accounts, this experience is being gained in Afghanistan.[88]

PROSPECTS

The Afghanistan mission is an important test of NATO's out-of-area capability. In a view of growing prevalence, Afghanistan exemplifies conditions in which "extreme belief systems, ...unstable and intolerant societies, strategic crime and the globalization of commodities and communications combine to create a multi-dimensional threat transcending geography, function, and capability."[89]

As previously suggested, the NATO allies have maintained a basic unity of purpose in Afghanistan. Their desire to stabilize the country and to prevent the return of a terrorist state has led to an ongoing general consensus that ISAF can help to build a state that is relatively stable and no longer a source of international terrorism.

On the other hand, the growing level of violence carried out by what some perceive to be a resurgent Taliban, reinforced by the a growing number of al Qaeda and other foreign fighters, and the perception that the Afghan government has not made tangible progress in extending its authority, could lead to some wavering among the allies with respect to a long term commitment to remain in Afghanistan. Many observers predict that ISAF's efforts to stabilize Afghanistan will require at least five more years, or longer. This prospect has exacerbated an existing problem that several allied governments already face with respect to declining support among their general populations. As the years wear on and the situation on the ground shows little progress and more violence, some policy-makers believe that the five-plus years time-frame could ultimately lead to new public pressure on NATO member parliaments to consider downgrading support, or at least reducing the military commitments to the Afghan mission. In a public opinion survey released by the German Marshall Fund in September 2008, only 43% of those Europeans polled supported combat operations against the Taliban.[90] As was pointed out by the Atlantic Council in its March 2008 Issue Brief, "the stalemate in Afghanistan poses a great dilemma for NATO: how can the 26 NATO governments convince their public to support a long-term effort in Afghanistan without clear indications of real progress either in the security or reconstruction sectors. Those allies with substantial forces fighting in Afghanistan are already fatigued by the political battles at home, as adverse domestic opinion challenges the governments..".[91]

Some observers believe a good portion of the negative public opinion within some allied nations could be directly attributable to an overall negative opinion of the foreign policy of the previous Bush Administration, especially its Iraq policy. These observers suggest that the new U.S. Administration could have a more positive effect on the international stage and could serve to help reverse some of the prevailing skepticism. For some observers, however, a renewed emphasis on Afghanistan by the Obama Administration could result in increased pressure on the NATO allies to send more troops to Afghanistan and lift their restrictive national caveats. They argue that a more secure environment will allow a more rapid development of the Afghan infrastructure, the economy and government institutions such as the military, police and judiciary. On the other hand, some believe that increased military engagement could lead to more combat operations, more violence and more casualties, a prospect many NATO allies would be reluctant to have to explain to their public. Should the new Administration fail to secure even modest new troop commitments from the allies and does eventually send upwards of 30,000 additional U.S. soldiers, some believe the Alliance will face the dual problem of their publics no longer willing to support the "American" war in Afghanistan and the U.S. public turning a sour view toward Europe and NATO.

NATO's exit strategy for Afghanistan requires supporting the development of the economic foundations of the country and providing the security for a fledgling government to find a stable political footing that excludes violence, reduces corruption, and creates a climate conducive to representative institutions. External factors will affect realization of this exit strategy. Stabilization of Afghanistan is closely linked to developments in and the intentions of neighboring Iran and Pakistan, a situation that many in the alliance believe demands a continuing U.S. presence.[92]

U.S. leadership of the alliance appears to be at a key moment. The allies believe that the success of the mission will also be a test of the United States' ability and commitment to lead NATO, even if they do not always agree with every element of U.S. policy in the country. The United States and its NATO allies have greater unity of purpose in Afghanistan for now. The ultimate outcome of NATO's effort to stabilize Afghanistan and U.S. leadership of that effort may well affect the cohesiveness of the alliance and Washington's ability to shape NATO's future.

Source: Map Resources. Adapted by CRS. (K.Yancey 11/22/05)

Figure 1. Map of Afghanistan

AUTHOR CONTACT INFORMATION

Vincent Morelli
Section Research Manager
vmorelli@crs.loc.gov, 7-8051

Paul Belkin
Analyst in European Affairs
pbelkin@crs.loc.gov, 7-0220

End Notes

[1] The original version of this Report was produced by former CRS analyst, Paul Gallis, Specialist in European Affairs.

[2] See CRS Report RS22633, *U.S. Forces in Afghanistan*, by JoAnne Bryan and Michael Waterhouse.

[3] Note: The number of "boots-on-the-ground" are approximations due to regular unit rotations and the different ways in which the U.S. Joint Staff and ISAF account for personnel.

[4] *Transatlantic Trends*, The German Marshall Fund, September 2008.

[5] "Gates asks Europeans to face Afghan threat," *International Herald Tribune*, Be. 9-10, 2008, p. 3.

[6] "Revitalizing our Efforts, Rethinking our Strategies," Report of the Afghanistan Study Group, Center for the Study of the Presidency, January 30, 2008, p.17.

[7] For an overview and analysis of key issues in Afghanistan, including the role of Pakistan, see CRS Report RL30588, *Afghanistan: Post-War Governance, Security, and U.S. Policy*, by Kenneth Katzman.

[8] "Pakistan will give Arms to Tribal Militias," *Washington Post*, October 23, 2008.

[9] Mazzetti, Mark and Eric Schmitt. "CIA Outlines Pakistan Links With Militants." New York Times, July 30, 2008.

[10] "ISAF's Strategic Vision," NATO summit, Bucharest, April 3, 2008, p. 1.

[11] Interviews with officials from European governments and U.S. specialists, April 2008.

[12] "NATO Commander Asks Member Nations to Drop Troop Limits," *Mideast Stars and Stripes*, October 25, 2006.

[13] Interviews at the NATO Defense College, Rome, December 2006, and Washington, DC, April-May 2007; "Germans wavering on Afghan mission," *International Herald Tribune*, August 20, 2007, p. 3.

[14] See Report, "Stabilization of Afghanistan: Challenges for the EU and International Community, Report (2007/2208(INI) of the Committee on Foreign Affairs, European Parliament, June 2008.

[15] Katzman, op. cit., p. 33

[16] Report of the Afghanistan Study Group, op. cit. p. 22

[17] Statement of Defense Secretary Robert Gates before the House Armed Services Committee, September 10, 2008.

[18] "Provincial Reconstruction Teams in Afghanistan - An Interagency Assessment," Dept. of Defense, Washington, DC, April 26, 2006; Interviews of U.S. officials, 2006-2008.

[19] Discussion with U.S. official, February 2008.

[20] Interviews of U.S. officials, 2005-2007.

[21] For example, see House Armed Services Subcommittee on Oversight and Investigations, hearing on PRTs, 1st session, 110th Congress, December 5, 2007.

[22] Testimony of Director Negroponte, "Annual Threat Assessment," Senate Select Committee on Intelligence, January 11, 2007; House International Relations Committee, hearing on "U.S. Counternarcotics Policy in Afghanistan," March 17, 2005; Mishra, *op. cit*, p. 46.

[23] Interview, June 20, 2007; and "Opium guerre, le 'narco-état afghan," *Le Monde* (December 13, 2007), p. 5.

[24] Interviews with officials from allied countries, June-December 2007.

[25] "Obstacles arise in Bid to Curb Afghan Trade in Narcotics," *New York Times,* December 23, 2008.

[26] "Foreign Troops in North Afghanistan Say 'Drug Wars' the Biggest Threat," *Agence France Presse*, August 30, 2005; "Shake-up of Afghan Police 'Brought Back Corruption,'" *Financial Times*, June 13, 2006, p. 2.

[27] Cited in "If Called to Lebanon, NATO 'Could Go In,'" *International Herald Tribune*, July 28, 2006, p. 3; interviews, fall 2006.

[28] "German giving up on Afghan position," *International Herald Tribune*, September 12, 2007, p. 1; interviews with officials from allied governments, June-September, 2007.

[29] "Quarterly Report to the U.N. on ISAF Operations," NATO, Brussels, February 1, 2008, p. 3.

[30] Interviews with European Union officials, 2006-2007; presentation of former Afghan Finance Minister Ashraf Ghani, Brookings Institution, April 30, 2007; and "McCaffrey Sees 2007 as a Crucial Year," *Washington Post*, April 10, 2007, p. A15.

[31] "Rome Conference on Justice and Rule of Law in Afghanistan," Rome, July 2-3, 2007; interviews of Italian officials, August 2007. The United States pledged $15 million for the program, and Italy pledged approximately $13.5 million.

[32] "Europeans Balking at New Afghan Role," *New York Times*, September 14, 2005, p. 1; interviews of European officials, September 2005 - February 2006.
[33] "Final Communiqué," North Atlantic Council, NATO, Brussels, December 8, 2005.
[34] "UK Warned of More Afghanistan Deaths," *Financial Times*, July 3, 2006, p. 3.
[35] "Europeans Balking at New Afghan Role," *op. cit.* Struck's view seems to be contradicted by the 1999 NATO Strategic Concept, the alliance's guiding political document, which clearly states that counter-terrorism is one of NATO's new post-Cold War tasks.
[36] "Peacekeeping in Afghanistan Is Modern Crisis Management," in *European Affairs*, spring/summer 2006, p. 3-4.
[37] Comments by Gen. Jones at NATO Parliamentary Assembly meetings in Copenhagen, November 2005.
[38] "Visit to Afghanistan," report by the Defence Committee of the NATO Parliamentary Assembly, May 23, 2006, p. 2.
[39] Testimony of Sec. Gates, House Armed Services Committee, hearing, 1st Session, 110th Congress, December 11, 2007.
[40] Interviews, June-December, 2007.
[41] *Transatlantic Trends, op. cit.*, p. 17-18.
[42] Interviews with NATO officials, February 2006.
[43] Interviews with officials from NATO governments, December 2005-February 2006; "En Afghanistan, l'OTANévolue de la pacification vers le contre-terrorisme," *Le Monde*, November 20-21, 2005, p. 4.
[44] "Final Communiqué," North Atlantic Council Ministerial meeting, December 8, 2005.
[45] Interviews with officials from NATO governments, 2005-2007.
[46] Interviews with officials from allied governments, December 2005-October 2007.
[47] Testimony of U.S. Secretary of Defense Robert Gates before the Senate Armed Services Committee, September 23, 2008.
[48] Katzman, op. cit., p.24.
[49] Interviews with European and U.S. officials and observers, June-July 2006.
[50] "Germany Plans to Raise Troops Level in Afghanistan," Spiegelonline, June 24, 2008.
[51] "Germany to Send More Aid to Afghanistan via UN Fund," *DPA News Agency, Deutsche Welle,* December 16, 2008.
[52] *Transatlantic Trends*, Key Findings 2008, the German Marshall Fund annual survey, September 2008.
[53] Judy Dempsey, "Merkel aloof as public wavers on Afghanistan," *International Herald Tribune*, October 19, 2007, p. 2.
[54] Discussions with Dutch officials, September 2005-May 2006.
[55] Bernard Bot, "Saving Democracy in a World of Change," speech at Georgetown University, Washington, DC, October 24, 2006; interviews, 2007.
[56] Remarks by Bert Koenders, Minister for Development and Cooperation, The Netherlands, at CSIS, Washington, DC, April 16, 2007. Koenders is the highly regarded former President of the NATO Parliamentary Assembly, and is well-versed in NATO issues. For a view advocating EU coordination of reconstruction/civilian programs in Afghanistan, see Julianne Smith, "How the EU Can Act Now to Assist Global Leadership," CSIS report, March 26, 2007.
[57] Remarks of Gen. Eikenberry at Brookings conference on Europe, April 30, 2007.
[58] Discussions with Dutch and U.S. officials, February-July 2006.
[59] Discussions with officials from the Dept. of Defense, October-December 2007; "An Afghan mission short of troops," *International Herald Tribune*, May 3-4, 2008, p. 3.
[60] Discussions with U.S. officials, 2006-2007.
[61] "NATO plans to reduce Afghan casualties," *Financial Times*, July 30, 2007, p. 1.
[62] "For British in an Afghan province, initial gains against the Taliban,"*New York Times*, August 5, 2007, p. 1.
[63] "Provincial Reconstruction Teams ...," *op. cit.*, Dept. of Defense, p. 22; "Opium War an Absolute Disaster," *Financial Times*, July 5, 2006, p. 3.
[64] "Fields of little glory," *Financial Times*, November 19, 2007, p. 10.
[65] "Canadian and Dutch Publics Feeling Stretched ...," op. cit.; "Troop Pullout Bill Defeated in Canada," *Washington Post*, April 25, 2007, p. A12.
[66] Interview with Defense Minister Morin in "Hervé Morin: 'La situation se dégrade en Afghanistan," *Le Monde*, December 21, 2007, p. 5.
[67] "La France redéploie ses avions de combat dans le Sud afghan," *Le Monde*, August 31, 2007, p. 4.
[68] NATO quarterly report to the U.N., op. cit.; "Mille soldats français en renfort dans l'Est afghan," *Le Monde*, April 2, 2008, p. 4.
[69] "L'Opposition français craint un 'enlisement' en Afghanistan," *Le Monde*, April 3, 2008, p. 6; and "A Kaboul, M. Sarkozy évoque un effort militaire français," *Le Monde*, December 25, 2007, p. 4.
[70] "À Kandahar, dans la base sous haute sécurité, d'où opèrent les Mirages français en Afghanistan," *Le Monde*, November 25-26, 2007, p. 5.
[71] Interviews with French and U.S. officials; Remarks by Defense Minister Michèle Alliot-Marie at the NATO Parliamentary Assembly plenary, Paris, May 30, 2006.

[72] CRS Report RS21659, *NATO's Prague Capabilities Commitment*, by Carl Ek.

[73] Interviews with U.S. and French officials, 2005-2007; "France Quietly Offers More Military Help," *Army Times*, August 29, 2005; "Français et Américains louent une coopération exemplaire en Afghanistan," *Le Monde*, October 24-25, 2004, p. 3.

[74] Interviews with French officials, August 2005-July 2006; Alliot-Marie, *op. cit.* Afghanistan supplies an estimated 90% of the heroin that finds its way to France; "Hervé Morin: 'La situation...," op. cit.

[75] See CRS Report RL34276, *FY2008 Emergency Supplemental Appropriations for International Affairs*, by Susan B. Epstein, Rhoda Margesson, and Curt Tarnoff.

[76] For a detailed description, see CRS Report RL34278, *FY2008 Supplemental Appropriations for Global War on Terror Military Operations, International Affairs, and Other Purposes*, by Stephen Daggett et al..

[77] National Defense Authorization Act for Fiscal Year 2008.

[78] "Saving Afghanistan: An Appeal and Plan for Urgent Action," Issue Brief, The Atlantic Council of the United States, March 2008.

[79] House Armed Services Committee, hearing on "Security and Stability in Afghanistan," June 28, 2006.

[80] "UK Troops 'Must Beat Back the Taliban this Year,'" *Financial Times*, May 23, 2006, p. 7; interviews with U.S. and European officials, 2006-2007.

[81] "Mission Impossible? Why Stabilising Afghanistan Will Be a Stiff Test for NATO," *Financial Times*, July 31, 2006, p. 9. The quotation is a paraphrase by the *Financial Times* of a French official who was reflecting on a similar dilemma for French forces in Algeria in the 1950s.

[82] Interviews with U.S. and British officials, 2005-2007.

[83] "Army Woefully Unready, Afghans Say," *Globe and Mail*, November 16, 2006, p. A13.

[84] Ibid.

[85] House Armed Services Committee, June 28, 2006, *op. cit.*

[86] "EU/Afghanistan: Europeans must Prepare for Losses ...," *Atlantic News*, July 20, 2006, p. 2.

[87] "Mille soldats français...," op. cit.

[88] Interviews with military officers from NATO governments, 2006-2007.

[89] Julian Lindley-French, "Big World, Big Future, Big NATO," *NATO Review*, Winter 2005, p. 5.

[90] *Transatlantic Trends, op. cit.,* September 10, 2008.

[91] "Saving Afghanistan: An Appeal and a Plan for Urgent Action," op. cit., p.7

[92] Olivier Roy, "Afghanistan: La Difficile Reconstruction d'un État," *Cahiers de Chailliot*, December 2004.

In: War in Afghanistan: Strategy...
Editor: Easton H. Ussery, pp. 33-52
ISBN: 978-1-60741-579-4

Chapter 2

UNITY OF COMMAND IN AFGHANISTAN: A FORSAKEN PRINCIPLE OF WAR

Colonel Ian Hope
Princess Patricia's Canadian Light Infantry

ABSTRACT

This Carlisle Paper discusses the traditional importance of unity of command in American doctrine and practice from World War I until now, and how this principle has been forsaken in the evolution of military command for Afghanistan. It examines the unprecedented departure from the principle of unity of command in Afghanistan in 2006, when Combined Forces Command-Afghanistan passed control of the ground fight to the International Security Assistance Force, and operations became split between several unified or "supreme" commanders in charge of U.S. Central Command, the North Atlantic Treaty Organization, and U.S. Special Operations Command. It argues for a renewal of understanding of the importance of unity of command, and recommends that the United States revert to the application of this principle by amending the Unified Command Plan to invest one "supreme commander" with responsibility for the current Operation ENDURING FREEDOM Joint Operations Area.

Unity of Command: Unity of command is best achieved by vesting a single commander with requisite authority.

—Principles of War 1954[1]

In Afghanistan today, want of moral singleness, simplicity, and intensity of purpose harp of military failure. This is attributable to an abrupt departure from a long-standing and distinctly American practice of insisting on unity of command. The United States is the only country where military doctrine recognizes the principle of "unity of command,"[2] and has successfully applied it in multiple alliances and coalitions since 1918. It was the guiding principle during World War II that convinced Allied powers to invest "supreme command" upon singular operational level commanders in distinct geographic areas.[3] Unity of command was the principle behind the 1946 Unified Command Plan (UCP), which institutionalized the practice of unifying forces under one commander-in-chief. This paper examines the departure from this principle that occurred in Afghanistan in 2006, when Combined Forces Command-Afghanistan (CFC-A) passed control of the ground fight to the International Security Assistance Force (ISAF), and operations became split between Commander U.S. Central Command (CENTCOM), Supreme Allied Commander Europe (SACEUR), and Commander U.S. Special Operations Command (SOCOM).

The paper has three main parts. Part one defines "unity of command" and describes how the United States has tried to adhere to this principle since 1914. It contains a brief synopsis of the American experience of coalition warfare in both World Wars, and reviews the evolution of "unified command" as a continuous attempt to reconcile geographic, coalition, functional, and service differences. Part two focuses on the evolution of the command structure in Afghanistan since 2001. The third part analyzes the current command structure in Afghanistan. Historical documentary evidence is used here to analyze the divisive roles played by CENTCOM, the North Atlantic Treaty Organization (NATO), SOCOM, and even the Department of State (DOS) and U.S. Strategic Command (STRATCOM), illustrating six areas where traditional unity of command has not been properly applied.[4] Part three also provides a recommendation to revive unity of command in Afghanistan through an amendment of the UCP.[5]

THE AMERICAN TRADITION OF UNITY OF COMMAND

American practice of unity of command requires the placement of *all* forces operating in a specific theatre to achieve a distinct objective under a single commander. This originated in the Civil War, and reached an apogee with the ascension of General U. S. Grant as General-in-Chief of the U.S. Army; an investiture of supreme command designed to unify all northern military efforts under one brain.[6] By 1914, the idea had become a "Combat Principle," articulated in *Field Service Regulations* as such: "Unity of command is essential to success . . . . All troops assigned to the execution of a distinct tactical task must be placed under one command."[7] It took the desperate situation created by the German offensive of March 1918 before allied generals could accept this essential principle. The Supreme War Council granted General Ferdinand Foch "supreme command" over French, American, and British Imperial forces on March 25. This finally provided unprecedented singleness of purpose in allied planning. Foch's July 24 memorandum became the blueprint for the coordinated offensives that defeated Germany that autumn.[8]

But unity of military effort did not equate to abandonment of complete sovereignty of American or British Imperial forces to French design. The relationship between Generals John Pershing and Foch was often strained due to the caveats placed upon the use of American forces. As leader of an "Associated Power," Pershing had Presidential mandate to resist piecemeal engagement of the American Expeditionary Forces. Pershing wanted to wait until an independent U.S. Army was ready to take the field (1919). Allied commanders desperately wanted American manpower deployed to their portion of the front, but acquiesced to Pershing's overall demand to retain national identity of American forces. With mutual concession, Pershing recognized the crisis and granted the employment of U.S. divisions under French command in mid-1918.[9] (See Figure 1.)

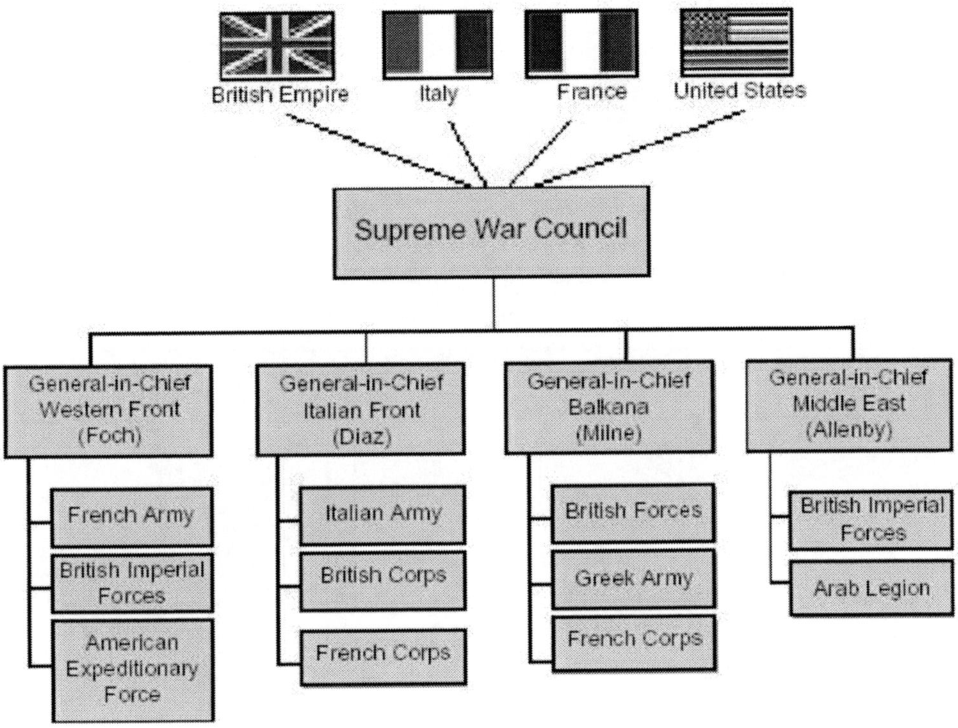

Figure 1. Unity of Command 1918.

At the end of the Great War, allied senior leadership recognized that singular command over multinational forces, even allowing recourse to national authorities, was the sole means to achieve the operational level cognition and cohesion essential to unity of effort. Colonel T. Bentley Mott, Pershing's Liaison officer at Foch's combined headquarters, remembers presenting Foch with General Pershing's concerns during the Argonne offensive. Foch whisked him to a billiard table covered with western front maps and said: "I am the leader of an orchestra. Here are the English Bassos, here the American baritones, and there the French tenors. When I raise my baton, every man must play, or else he must not come to my concert."[10] Foch spoke with authority invested by political agreement, and his "supreme command" represented a singular unified military operational effort under a mutually accepted alliance strategy. While often at odds with his "orchestra leader," Pershing later reflected: "I do not believe it is possible to have unity of action without a supreme commander."[11] The lesson was not lost to junior observers.

General George C. Marshall had been Pershing's Chief of Operations and had wit-nessed his commander's resistance to subordination under Foch. He had also witnes-sed the positive effect of Foch's "strategic direction."[12] Upon assuming responsibilities as U.S. Army Chief of Staff in 1939, Marshall immediately began to shape the environ-ment in Washington to embrace the principle of unity of military command under civil authority. In this, he relied upon U.S. Army doctrine formulated and instructed during the inter-war period.[13]

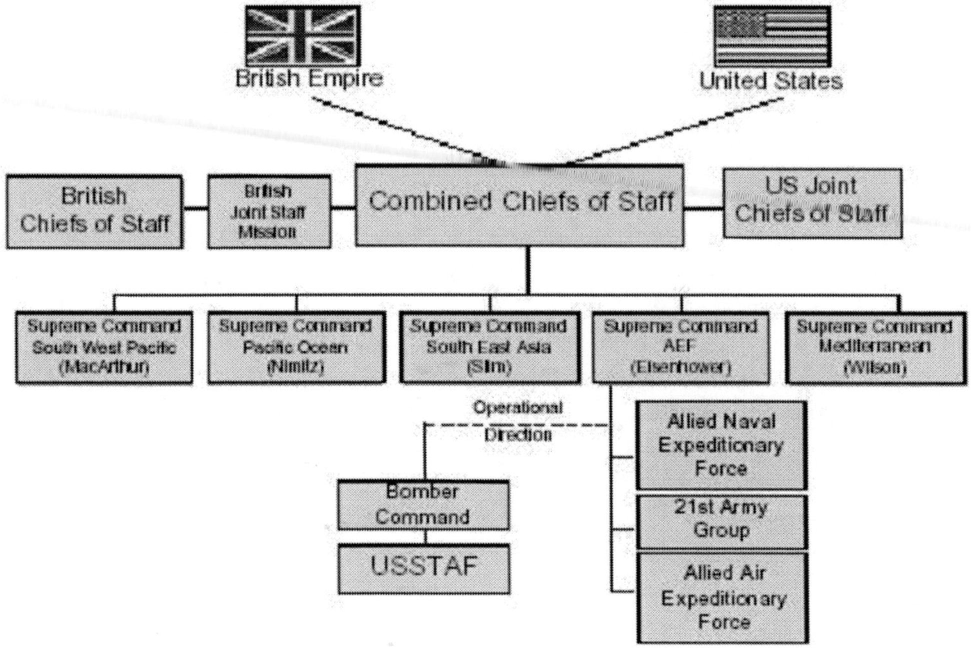

Figure 2. CCS and Supreme Commands 1945.

Shortly after the December 1941 attack on Pearl Harbor, Marshall concluded that "unified command . . . would solve nine-tenths of the problems of British-American military collaboration"[14] Thereafter he worked tirelessly to establish unified commands in each major theater of the war, uniting all services of every participating nation under one commander-in-chief (CINC).[15] Simultaneously, Marshall addressed the prospect of debilitating competition between CINCs, whose individual horizons were too narrow to appreciate the larger war-management problem.[16] He forced the U.S. military to adapt to the British committee system in managing the strategy of the war,[17] in which he and Admiral Ernest J. King became part of the British-American Combined Chiefs of Staff (CCS), holding unquestioned authority over the theater commanders. The CCS functioned through a system of standing committees and programmed meetings where the service chiefs and political leaders of leading alliance nations met to determine the course of the war. As such, it addressed "grand strategy" where national war aims were amalgamated to produce sanctioned military strategy to be implemented by theater commanders.[18] (See Figure 2.)

The CCS system also helped to overcome another obstacle to unity of command—service rivalry. Marshall contended with bitter interservice fights in the process of determining theater command constructs and strategic objectives in the Pacific, where the U.S. Army and the U.S. Navy held unbending and competing ideas of strategy and priorities.[19] These rivalries led to division of the Pacific theater into service-oriented subtheaters; the Navy-dominated Pacific Ocean Area under Admiral Chester Nimitz and the Army-dominated Southwest Pacific under General Douglas MacArthur. However, this compromise garnered agreement for the overall global unified command structure. This structure was formally represented in a study conducted by General Dwight Eisenhower for the U.S. Joint Chiefs of Staff (JCS) in early 1942 in which he divided the world into three major theaters of operations—the Pacific becoming an American area of responsibility (AOR); the Near and Middle East coming under British command; and Europe, the Mediterranean, and the Atlantic being shared between the United States and Great Britain.[20]

To achieve success within the CCS committee system, all parties had to relinquish a degree of sovereignty. It was understood that in the ways, ends, and means debate, everyone had to give a little. Initially it was the United States that did so in larger measure.[21] However, as time went on and as U.S. forces became the largest national entities within theaters, U.S. desires prevailed. Marshall was successful in harboring resources for the invasion of Europe despite British attempts to boost efforts in the Mediterranean, and put his chosen man—General Eisenhower—into the role of supreme commander Europe.[22] Eisenhower's character was well-suited for this coalition command, particularly in his ability to deal with his multiple political masters. This was important because the American experience of coalition warfare throughout World War II was one of constant subordination of military planning to political policy. Eisenhower answered to the CCS and frequently to issues raised by Winston Churchill. This appeared unusual to Eisenhower, but it reflected the need to handle political concerns from alliance partners whose finite resources could not be squandered, and was a necessary element of supreme command within an alliance.[23] While it had taken several years to determine workable command relationships within the alliance, once they were established within the unified command structures, the result was enduring moral singleness and unity of purpose that led to the defeat of the Axis powers and Japan.

Marshall and Eisenhower affirmed two lessons by war's end—the efficacy of a single CINC as essential to achieving military unity of effort in a given theater of war, and the requirement that theater commanders be responsive to a higher strategic body where the competing requirements of policy and military strategy come together to be debated. Experience in both world wars informed them that the advantage of military unity under singular geographic commanders was not in itself sufficient to sustain alliances; strategy formulation required compromises and pluralism beyond the capacity of a theater CINC. The first lesson was foundational in creating the UCP in 1946; the second guided the formation of NATO in 1949.

The UCP was established in the U.S. military to institutionalize joint theater command. It aimed to achieve unity of military effort by reducing the service rivalries that had characterized U.S. strategy formulation during the war.[24] It helped simplify the command and control of U.S. Army, Navy, and Air forces in designated areas by placing them under a single CINC, assisted by a joint staff ". . . with appropriate members from the various components of the Services under command in key positions of responsibility"[25] answering to the JCS. Initially seven geographic unified commands were designated: Far East Command (FECOM), Pacific Command (PACOM), Alaskan Command, Northeast Command, Atlantic Fleet, Caribbean Command, and European Command (EUCOM). Because unified commanders were not entitled to perform unique-to-service functions—administration, training, supply, expenditure of appropriated funds, or construction—these tasks could not be performed in the unified headquarters and had to be delegated to subordinate component headquarters. Therefore, while the newly designated unified commanders had authority to execute joint and combined operations, the sustainment of these operations remained squarely in the hands of the JCS, because only they could manage overall global war strategy and set priorities between unified commands. Service influence was also exercised within the UCP through the practice of affiliating each unified command with a parent service chief of staff, guaranteeing that the U.S. Navy dominated PACOM and the U.S. Army EUCOM indefinitely.[26]

The first proponent of the UCP was the U.S. Navy, which remained dissatisfied with the divided command of the Pacific in 1945, particularly after PACOM army forces were allocated to MacArthur for the occupation of Japan. With the UCP, Admiral Nimitz's Pacific Ocean Area Headquarters quickly transformed into Headquarters of the Commander-in-Chief Pacific Command (CINCPAC), and was reassigned army units. However, rivalry between the services continued. USPACOM maintained a long-standing competitive relationship with FECOM, where the U.S. Army ran Japan, Korea, the Ryukyus, the Philippines, the Marianas, and the Bonins.[27] (See Figure 3.)

The importance placed upon these arbitrary geographic boundaries and staunch service affiliations were conditioning factors that later had a negative impact upon the conduct of the Vietnam War. Early in that conflict, CINCPAC designated Commander U.S. Military Adviser Command Vietnam (COMUSMACV) as a subordinate unified commander within PACOM, but tightly circumscribed his responsibilities to land operations inside South Vietnam. While the war expanded, MACV's responsibilities did not, and control over the air, sea, and special operations missions outside the borders of South Vietnam, and many nonmilitary operations inside of South Vietnam, remained under control of CINCPAC, Commander-in-Chief Strategic Air Command (CINCSAC), and others. COMUSMACV lacked command authority over all the forces working within the contested area, and because he was required to work all joint and single service issues through PACOM, he lacked agency in Washington. This led the U.S. Army to request the establishment of Southeast Asia as a separate unified command or, alternatively, to secure the assignment of General Creighton Abrams to the position of CINCPAC upon completion of his duties as COMUSMACV. The U.S. Navy vetoed these proposals.[28] With failure in Vietnam, the limits of the UCP in dealing with the requirements of a protracted, complex, full spectrum operation became evident. Unity of command was not achieved during that conflict, costing the United States dearly. As a consequence, interservice rivalries and tensions between unified commands and service chiefs were studied extensively. The 1970 Blue Ribbon Panel reported to the President that unified commands had "too broad a span of control" and were "excessively layered, unwieldy, and unworkable in crises, and too fragmented to provide the best potential for coordinated response to a general war situation."[29] The observations were ignored in Washington until they were reviewed again under the Goldwater-Nichols' initiative in 1986.

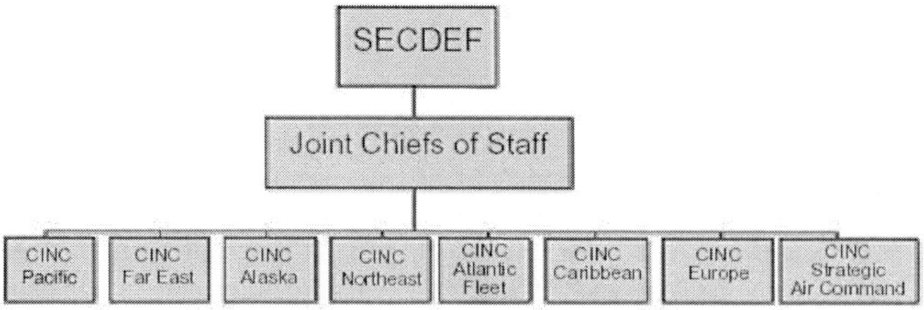

Figure 3. Unified Commands in the UCP, 1950.

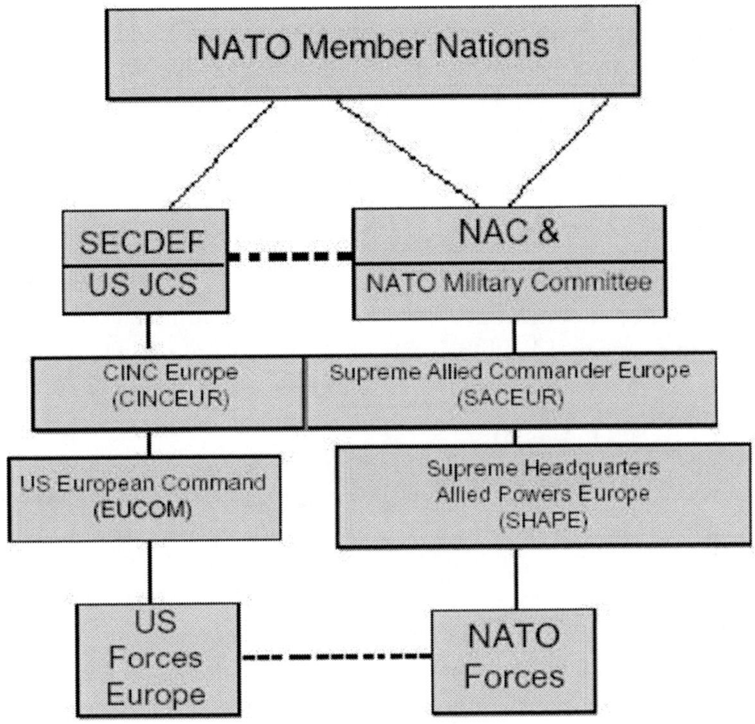

Figure 4. Unity of Command in NATO.

The other important geographic unified command was in Europe. Under the UCP, the senior American military headquarters in occupied Germany evolved to become U.S. European Command (EUCOM). This was commanded by the Commander-in-Chief Europe (CINCEUR) who was responsible for all U.S. forces on the continent. In a brilliant move to ensure unity of military command among allied powers, the USCINCEUR was also designated as Supreme Allied Commander Europe (SACEUR), in command of all NATO forces.[30]

NATO had been established in 1949 when Eisenhower's former Supreme Headquarters Allied Expeditionary Forces (SHAEF) transformed into Supreme Headquarters Allied Powers Europe (SHAPE). By that time, during European reconstruction and global decolonization, there was no questioning of American-led command constructs. Within the alliance, however, American military and political leaders understood that military command of a theater did not equate with American control of strategy formulation. This had to be done in a multilateral forum that could achieve results similar to the CCS committee system of World War II. Therefore, the NATO Military Committee was established, comprised of military representatives of each member; and the North Atlantic Council (NAC) was created, comprised of permanent representatives—or on occasion member states' key leaders. The NAC became the civil-political forum for debate over combined strategy to which SHAPE was responsive. Within SHAPE, and in the military chain of command in each member nation, the sentiment favoring powerful supreme command was so strong that it became entrenched. Europeans have always since deferred to SACEUR, provided that alliance strategy formulation remains in the Military Committee and the NAC. United Nations (UN)-

sponsored operations in the Former Yugoslavia in 1996 demonstrated how NATO operations under the UCP could be executed maintaining the principle of military unity of command under multilateral political oversight. SACEUR remained the singular operational level commander, responsive to both the NAC and to the Secretary of Defense (SECDEF). SACEUR accepted such frustrating factors as national caveats to sustain the alliance political support that has endured for over a decade. Operations in Kosovo in 1999 were far more problematic as lack of political agreement to attack Serbia created unprecedented caveats, yet SACEUR still maintained that: "The NATO process worked, . . . I was persuaded of the basic soundness of NATO decisionmaking."[31] Such conviction surprised many officers on the EUCOM staff who remembered how their CINC's orders were not always obeyed, and who sensed a division in operational purpose that Eisenhower and Pershing would have recognized as normal within an alliance, but was constraining to those working within the parallel U.S. unified headquarters.[32] They sensed a stark contrast between the constraints of the alliance and the success achieved by the ad hoc coalition through which CENTCOM executed operations in Saudi Arabia and Kuwait in 1990 and 1991. (See Figure 4.)

The first big test of the UCP in CENTCOM was Operations DESERT SHIELD/DESERT STORM. The CENTCOM commander, General Norman Schwarzkopf, approached the problem of coalition warfare understanding the dual requirement of military unity of command and a multilateral mechanism for strategy formulation. He wisely formed two coalition constructs, the first consisting of U.S.-led western troop-contributing nations, the second consisting of Saudi-led Arab troop-contributing nations. The Headquarters for both coalitions were collocated and managed through a Coalition Coordination and Communication Integration Center.[33] Officially co-commanded by General Schwarzkopf and Lieutenant-General Prince Khalid Bin Sultan al-Saud, the technical functions of supreme command were exercised by the CENTCOM commander. In this construct Schwarzkopf operated like Eisenhower, assuring both his U.S. masters and European and Arab political leaders that military action was circumscribed by coalition policy, acquiescing to the JCS and national command authorities who engaged international partners to craft strategy. Schwarzkopf was left alone to function as the singular "combatant commander," and was the center point that ensured singularity of purpose, and simplicity in structure of command. It is reasonable to say that CENTCOM's conduct of the operations was a realization of the intent of the UCP. The short duration of the war and its limited objectives granted a commonality of moral purpose that allowed unity of effort. Whatever problems that might have existed were also short-lived.

In several important aspects CENTCOM's success was attributable to UCP practices before Goldwater-Nichols. The UCP has evolved constantly since 1946, adding and collapsing geographic commands, adding, renaming, and amalgamating functional commands, and continuously trying to reconcile tensions between the commands and the services. However, the narrative history of the UCP is one of a very slow but steady trend toward increasing the unified commanders' authority vis-à-vis the service chiefs', a relationship at the center of the Goldwater-Nichols Act of 1986. The Act enhanced the agency of the unified commanders by allowing them to report directly to the SECDEF. While this may be seen as reinforcing the principle of military unity of command, it is actually a significant challenge to the original UCP construct as it marginalized the JCS and set up the conditions where unified commanders and service chiefs would compete. By placing the combatant commanders squarely under the SECDEF, Goldwater-Nichols produced the command arrangement that was desired and achieved by Pershing in 1917, but was opposed by Marshall throughout World War II. It eroded the ability of the U.S. military's strategic echelon to participate in strategy formulation and make the hard choices regarding ways, ends, and means between competing theaters, and has obstructed JCS' efforts to work with allies to determine combined objectives in war management.

Ironically, the problems associated with Goldwater-Nichols remained masked by military success in Operation DESERT STORM. This, coupled with the remembered frustration of EUCOM in Kosovo in 1999, shaped U.S. thinking about coalition warfare in the aftermath of the tragic events of September 11, 2001 (9/11). Sentiment favored coalitions (defined by U.S. military missions) over alliances with constraining forums like the Supreme War Council, the Imperial Chiefs of Staff, the CCS committee system, or the NAC. In the post 9/11 world, war strategy would be confined within the Executive Branch of the U.S. Government, and executed by U.S. combatant commanders, with tactical support from "invited" coalition members. The problems this created surfaced in Afghanistan.

Coalition Unity of Command in Afghanistan

At the commencement of Operation ENDURING FREEDOM (OEF) in 2001, "supreme command" fell upon Commander CENTCOM who decided not to create a subordinate unified command in Afghanistan. In the absence of a combined strategic forum, CENTCOM also assumed lead role in coalition war management: a huge departure from past practices. Sympathy for the United States, and the assumption that operations in Afghanistan would be short, caused few nations to raise political objections to a CENTCOM lead, even though CINCCENT[34] was not subject to any non-U.S. political scrutiny. Nor did CINCCENT feel compelled to subject himself to any political concerns other than those of the SECDEF or the President. Aside from guidance emanating from the Bonn conference, and from bilateral military-to-military relationships, war strategy for Afghanistan was made almost entirely in Washington, DC, and Tampa, Florida (despite the fact that the Bonn Process allowed four other nations—Germany, Italy, the United Kingdom [UK], and Japan—to work individual initiatives for police, justice, counternarcotics, and disarmament reforms in Afghanistan independent of CENTCOM or the SECDEF). From a purely army-centric perspective, non-U.S. coalition contingents were "unified" under one CINC in Tampa, where each coalition member had a liaison team. From a strategic perspective, no one was in charge of the overall Afghanistan mission.

In 2001, CINCCENT deployed a CFLCC, a CFACC, and a CFMCC to the Arabian Gulf, and a special operations task force (now called Combined Joint Special Operations Task Force [CJSOTF]) into Afghanistan.[35] However, unifying the tactical efforts of these multiple service components and their coalition partners was very difficult, as evidenced in Operation ANACONDA in March 2002.[36] Therefore, in 2003, CFC-A was established as the joint operational level headquarters for Afghanistan. A subordinate unified command, CFC-A was also responsible for building the Afghan Army (through the Office of Military Cooperation—now called Combined Security Transition Command Afghanistan [CSTC-A]), pushing reconstruction efforts through their newly established Provincial Reconstruction Teams (PRT), and managing the joint special operations fight.[37]

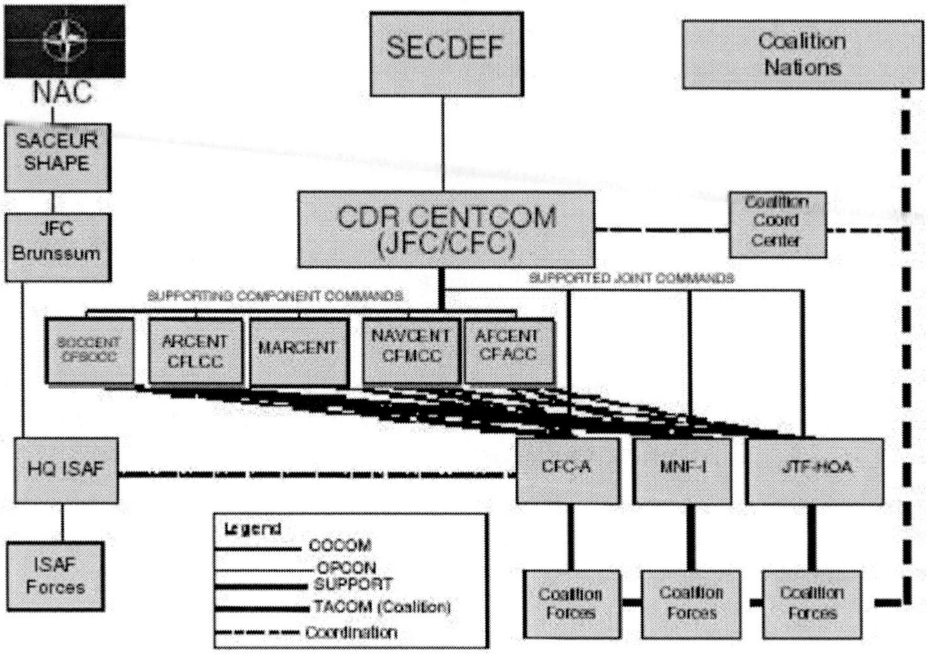

Figure 5. Command Relationships, CENTCOM 2003.

CFC-A was supported by the CENTCOM CFLCC, CFACC, and CFMCC, who were also supporting Operation IRAQI FREEDOM (OIF). Under this construct, tactical unity of command existed for in-country ground combat; however, the CFC-A commander quickly found himself competing for critical ground, close air support (CAS), and Intelligence/Surveillance/Reconnaissance (ISR) assets against the OIF mission. While the CFLCC, CFACC, and CFMCC considered him "supported commander," he was not the primary supported commander in CENTCOM, and his theater became an economy-of-effort mission to operations in Iraq. Somewhat akin to the problem faced by COMUSMACV, the commander of CFC-A found that his agency was limited by having to work his Title 10 concerns through a CFLCC and a CFACC serving dual missions, and his theater concerns through a combatant commander preoccupied with other operations. (See Figure 5.)

Unity of command and political oversight of multinational forces emerged as issues with the growth of ISAF. Originally an independent, UN-mandated, British-led mission overseeing the post-conflict transition of Kabul, ISAF had no command relationship with the senior U.S. headquarters in Bagram.[38] American staffs believed that ISAF was tactical control (TACON) to the CENTCOM CLFCC, but in reality ISAF worked through national channels to Britain and coordinated non-British assets through coalition representatives in Tampa.[39] This independent approach was confusing, and it ended in 2003 when the ISAF mission was taken over by NATO and command and control moved to an entirely European chain of command from ISAF HQ in Kabul, Afghanistan, to NATO Joint Forces Command (JFC)-Brunssum in the Netherlands (commanded by a German four-star general), then upward to SHAPE. NATO assumption, however, did not clarify a relationship between the ISAF and OEF missions.

ISAF HQ changed every 6 months as designated NATO corps headquarters assumed the mission on a rotating basis, ensuring no continuity in command and little progress in establishing a standing relationship with HQ CFC-A. As the senior American headquarters in Afghanistan, CFC-A saw itself as the superior headquarters, functioning at the operational level.[40] But from the NATO perspective, CFC-A was another tactical level headquarters, separate and distinct from the three-star ISAF tactical headquarters, and certainly not its superior.[41] Unable to see or reconcile this difference in perspectives, and each side assuming it was correct, a decision was nonetheless made to expand the ISAF area beginning in 2004 to assume responsibility for the northern part of the country, the western part of Afghanistan in 2005, and all territory in Afghanistan in late 2006.

This decision was predicated upon the U.S. desire to reduce its military commitment to Afghanistan—backfilling American troops with NATO forces, and migrating U.S. functions to EUCOM.[42] The final and most sensitive part of this expansion (bringing Regional Command South and Regional Command East under NATO) occurred between May 2006 and February 2007. At this time, the British and American-dominated NATO Allied Rapid Reaction Corps Headquarters assumed the role of HQ ISAF, and a Canadian-led Brigade accepted the task of working under OEF to oversee NATO deployment into Regional Command-South, allowing for its transfer to ISAF in August 2006. Once that was successfully accomplished, U.S. ground forces operating under OEF in Regional Command-East were placed under NATO/ISAF in November.

SHAPE looked to JFC Brunssum as the operational level headquarters under which ISAF was a tactical component. Yet Brunssum was ill-equipped for this task, and was too far removed from the realities of Afghanistan to provide the necessary planning or operational guidance. Most importantly, JFC-Brunssum had no authority over the U.S. headquarters remaining in Afghanistan, making unity of command impossible. In a benign stability operation, this might have been made to work, but events in 2006 emphasized the divisions. NATO found itself inheriting a growing insurgency they had previously dismissed as an American problem. Fighting produced a growing number of Canadian and British casualties, and NATO was unprepared psychologically for this development. NATO member nations had not been socialized for the combat requirements of Afghanistan, and balked when this realization occurred.

Despite this, CFC-A was disbanded in February 2007, and ISAF headquarters became responsible for the country before the unity of command issue was addressed. Although NATO assumed responsibility, SACEUR/CINCEUR was still not the combatant com-mander that the United States held accountable for Afghanistan. He had no relationship with critical supporting U.S. headquarters—especially the CLFCC, CFACC, and CJSOTF, and CSTC-A.[43] These remained with CENTCOM. Through CSTC-A, CJSOTF, and the senior U.S. tactical headquarters in Afghanistan—at time of writing Task Force 82, soon to become Task Force 101—CINCCENT continues to exercise authority for six critical functions that historically would have been transferred to SACEUR/CINCEUR if unity of command was still important: (1) U.S. title 10 (including logistics and medical support) responsibilities; (2) capacity-building of the Afghan security forces; (3) special operations coordination; (4) ISR and CAS support, (5) counternarcotics coordination; and (6) regional engagement with neighboring countries, most importantly Pakistan.[44] NATO and EUCOM have no part in these efforts.

The mixing of command authorities inside the Afghanistan theater is a second major departure from 60 years of practice in the UCP. The first occurred in Vietnam.

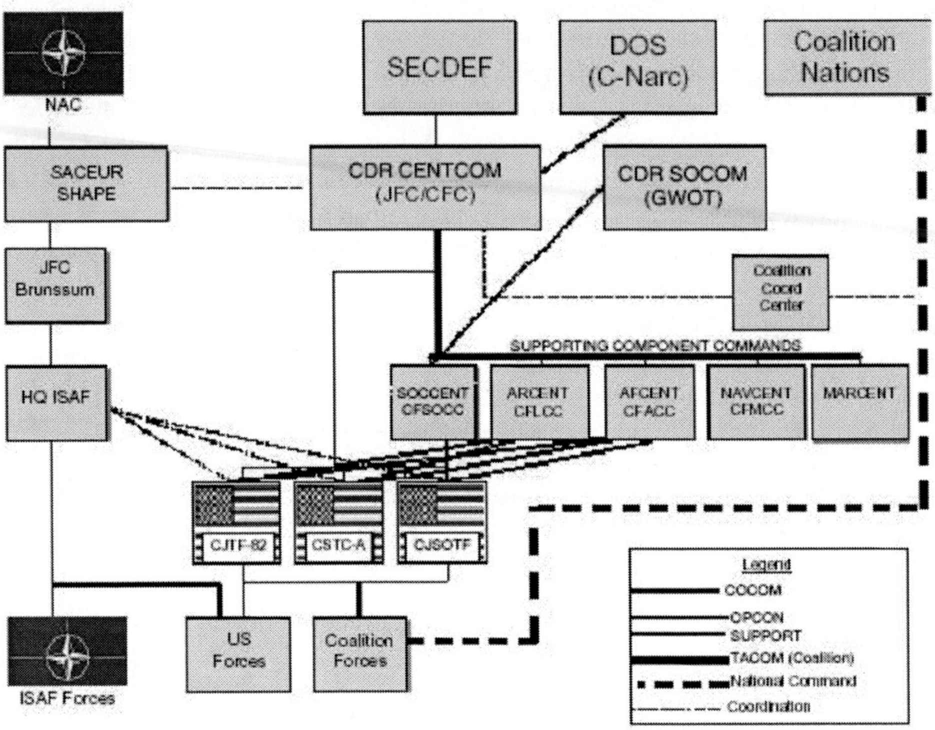

Figure 6. Command in Afghanistan 2008.

Analysis and Recommendation

While SACEUR's soldiers fight in Afghanistan, CINCCENT retains control of the U.S. service component contributions to the fight, including the CLFCC and CFACC, the CSTC-A functions, and development functions in the U.S. PRTs. CENTCOM works with SOCOM to coordinate the counterterrorist fight, and with the DOS to support counter-narcotics operations; and CINCCENT engages with Pakistan to coordinate counterinsurgency and counterterrorist operations. SACEUR has no involvement in these activities. The White House and CENTCOM have been reluctant to shift any of these functions to NATO and EUCOM because they fear being constrained by the alliance. At the same time, NATO members are suspicious of continued CENTCOM involvement, and have placed heavy caveats upon their forces to protect them from being sucked into OEF missions that are directed unilaterally by the White House and CENTCOM with no alliance input. U.S. reluctance to work within NATO and European refusal to support U.S. unilateralism have created a fractured command structure that is abetting the Taliban insurgency and the forces of corruption that plague Afghanistan.

Realignment of all U.S. functions under USEUCOM and empowerment of SACEUR/CINCEUR would solve all six of the issues currently fracturing command in Afghanistan. Officers in Washington and Tampa realize that the CENTCOM CLFCC and CFACC supporting Afghanistan and Iraq is highly efficient for controlling U.S. force rotations, logistics, medical support, ISR, and CAS allocations within the CENTCOM AOR. However, retention of these functions in CENTCOM makes full NATO integration and involvement in Afghanistan impossible. Unity of command would be better achieved by the establishment of a EUCOM/NATO JLFCC and JFACC, supporting a U.S./NATO JFC for Afghanistan. All U.S. Title 10 responsibilities would be fulfilled, with the added benefit of NATO alignment in securing troop contributions, coordinating troop rotations, integrating logistics, and reducing redundancies (benefits realized by the creation of Eisenhower's SHAEF headquarters in February 1944).[45]

CSTC-A's Afghan National Army (ANA) capacity-building function and CENTAF's Afghan Air Corps capacity-building initiative should be realigned to USEUCOM and integrated with NATO staffs.[46] They should directly support the commander ISAF, with headquarters ISAF and CSTC-A reporting to the same boss, SACEUR/CINCEUR. This would allow for needed synchronization of ANA training and fielding with ISAF operations, something that has been dysfunctional under current command arrangements. Forces under ISAF mandate have a problematic relationship with CSTC-A (under OEF mandate). The training, equipping, and fielding of ANA battalions (Kandaks), their integration into operations, and their continued mentoring and sustainment are essential elements to ISAF mission success. But ISAF controls nothing in the process. Each NATO operation relies upon ANA. Yet, the allocation of ANA Kandaks is controlled by the United States.[47] For the past 3 years, this has been largely disproportional, with U.S. SOF and U.S. conventional forces in Regional Command East getting the majority share. This has left NATO nations scrambling to get a bilateral commitment from the United States for Kandak partnerships. NATO members resent this.[48] While the United States has stated a desire for the "NATO-ization" of the CSTC-A mission, its Title 10 and Title 22 funding (NATO will not assume the two billion dollar price tag) necessitates that it have more U.S. national and less alliance accountability. But this could be served under EUCOM, allowing the integration of NATO money and personnel considerably easier and more palatable to Europeans.[49]

Realignment under U.S. EUCOM (with EUCOM JFLCC and JFACC) would require the establishment of a coalition air operations center (CAOC) for operations in the Afghanistan JOA. This would alleviate current problems associated with reliance upon a singular CENTCOM CFACC/CAOC that provides ISR and CAS for competing missions in Afghanistan and Iraq. In the current construct, ISAF is severely disenfranchised, as there is no formal relationship between NATO and the CENTCOM CFACC (NATO officers are not even allowed into Qatar to serve in the CAOC as there are no standing forces agreements between NATO and Qatar). Presently, ISAF must work diplomatically through U.S. officers to secure the CENTCOM assets. This creates a perception among NATO partners that they are not receiving proper allocation of this support. This is unfair, because CAS remains one of the most reliable assets in Afghanistan, but the perception remains that NATO efforts are second fiddle to OIF and OEF by virtue of a command structure that cannot alleviate the suspicion. Creation of a EUCOM/NATO CAOC would give the ISAF commander a supporting air component that would be responsive to the same chain of command (SACEUR/CINCEUR), and create unity of command so lacking today.

ISAF commanders work beside U.S. SOF daily, and there is significant mixing of all forces in certain areas of Afghanistan. However, that these forces operate under different mandates and report to different combatant commanders remains problematic. U.S. SOF are governed by OEF and GWOT missions that emphasize foreign internal defense (FID) and counterterrorism, with reporting through the CJSOTF to either CENTCOM or SOCOM. ISAF does not involve itself in FID or counterterrorism, but is fighting in a counterinsurgency role in the same geographic area. This superimposition of different missions and chains of command upon the same piece of terrain is problematic. Most often these superimposed operations are well-coordinated and executed. However, recurring friction is inevitable when ISAF troops unintentionally compromise a SOF mission, or when SOF missions produce adverse effects that impact negatively upon ISAF soldiers.[50] Unity of command is the central issue here, and the obvious improvement to be made is realignment of all special operations under a U.S./NATO CJSOTF responsible to EUCOM, who would coordinate with SOCOM. This would once again allow the ISAF commander to trust his boss—SACEUR/CINCEUR—to represent his concerns in Washington, while at the same time reduce allied suspicions about who was really in charge in Afghanistan.

Currently the CENTCOM JIATF attempts to coordinate with DOS to synchronize counternarcotics efforts in Afghanistan. Europeans remain frustrated with U.S. counternarcotics initiatives, and see the involvement of CENTCOM as indicative of a desire to continue U.S. unilateralism. The record high yields for opium production during the past 4 years attest to the failure of this effort, and reinforce European skepticism. It is difficult to argue that theater unity of command under SACEUR/CINCEUR will solve the issues of counternarcotics coordination; but centralization of the U.S. effort in Europe would certainly help. The opium problem is one of strategic import, and requires a unified strategy that can only be produced by a multilateral body that can formulate strategy and prioritize in a manner similar to the work of the CCS during World War II. It is an issue for the NAC. So, too, is the concern about how best to engage Pakistan and other regional neighbors. Currently, CINCCENT has the U.S. lead with regard to this critical function, and SACEUR, whose troops are bearing the brunt of perennial insurgent offenses from Pakistan, remains as hamstrung as COMUSMACV was in dealing with communist force incursions into South Vietnam. The problem is that the Taliban insurgency, like that of opium, has regional dimensions and requires multilateral strategy and commitments. These problems are by their nature long term, necessitating commitment of multinational resources for decades to come. The NAC has 60 years of verifiable success in dealing with these sorts of complex issues, and, once it passes through the frustrating process of strategy formulation, offers the advantage of an enduring alliance as the mechanism to ensure that such commitments can be sustained. Its longevity and its ability at formulating acceptable strategy give it a clear advantage over the ad hoc coalitions that are currently eroding in OEF, OIF, and the GWOT.

Solving the command problem in Afghanistan requires renewal of our understanding of the principle of unity of command.[51] It requires recognition of the wisdom of Pershing, Marshall, and Eisenhower, and of reaching again an appreciation for the importance of singularity of purpose, and simplicity that comes with investiture in a "supreme commander." It is, therefore, the conclusion of this research that we must amend the UCP and invest supreme command over Afghanistan in SACEUR. In order to galvanize NATO alliance partners and begin the difficult process of coalition building around a NATO-run fight, while keeping parallel American capabilities in-theater, the entire OEF joint operating area (JOA) must be realigned under EUCOM, and EUCOM must be designated as a supported combatant command. The ISAF Headquarters in Kabul should be designated as an integrated sub-unified command under EUCOM to report directly to SHAPE. Separate EUCOM CFLCC, JFACC, and JFSOCC must be established so that the Afghan fight can receive proper Title 10, air,

ISR, and SOF support without having to go to Tampa for arbitration over which major operation—OIF or OEF/ISAF—gets priority. While hardly efficient from an American perspective, it is the only way that NATO partners can be integrated to fight under their traditional supreme commander, and under the alliance's normal strategic war-management system.

CONCLUSION

This Carlisle Paper has discussed the traditional importance of unity of command in U.S. doctrine and practice from World War I until now, and how this principle has been forsaken in the evolution of command construct in Afghanistan. It has argued for a renewal of understanding of the importance of unity of command, and recommends that the United States revert back to application of this principle by amending the UCP and granting responsibility for the current OEF JOA to USEUCOM. This would see two immediate improvements. First, it would invest SACEUR with the "supreme" authority over operations in Afghanistan that he presently is denied; second, it would make full use of a long-standing alliance to ensure the formulation of strategy and the sustainment of commitment that is obviously missing in the region today. This realignment would require designation of EUCOM as a supported combatant command, establishing EUCOM/NATO JFLCC and JFACC, the embedding within ISAF HQ the necessary elements to create an integrated subordinate unified command in Kabul, and streamlining the chain of command to have HQ ISAF report directly to SHAPE and SACEUR. To ensure full unity of command, the United States should transfer its counternarcotics and regional engagement functions to the NAC and NATO military council, and consolidate Title 10 and special operations functions under EUCOM.

Failure to address the current problems of unity of command will result in the failure of the alliance—and the coalition—in Afghanistan. The threats posed by the large-scale and enduring cross-border insurgency, steadily growing opium production, and endemic corruption, are sufficient to defeat our bifurcated military and civilian efforts in that conflicted country. We should heed the words of Eisenhower:

> Alliances in the past have often done no more than to name the common foe, and "unity of command" has been a pious aspiration thinly disguising the national jealousies, ambitions and recriminations of high ranking officers, unwilling to subordinate themselves or their forces to a command of different nationality or different service. . . . I was determined, from the first, to do all in my power to make this a truly Allied Force, with real unity of command and centralization of administrative responsibility.[52]

End Notes

[1] U.S. Department of the Army, *Field Service Regulations–Operations, Field Manual 100-5*, Washington, DC: U.S. Department of the Army, 1954, pp. 25-27. Full Text:
Unity of Command. The decisive application of full combat power requires unity of command. Unity of command results in unity of effort by coordinated action of all forces toward a common

goal. Coordination may be achieved by direction or by cooperation. It is best achieved by vesting a single commander with requisite authority.

[2] French, Russian, and British lists of the principles of war never mention "unity of command." The closest foreign principles are the principle of "unity of effort" in French doctrinal circles, and the principle of "cooperation" in British doctrine. For the unchanging French principle of "unity of effort," see France, Minister of War, *Manual for Commanders of Infantry Platoons*, 1917, trans. in the U.S. Army War College, Washington, DC: U.S. Army War College, 1917, pp. 157-158; and Ferdinand Foch, *Principles of War*, London: Chapman & Hall, Ltd., 1920. The principles of war in British military doctrine have also remained constant for 80 years, with the principle of "cooperation" worded similar to the American "unity of command"; see Great Britain, War Office, *Field Service Regulations, Vol.* II, *Operations, Provisional,* London: War Office, *Field Service Regulations, Vol. II, Operations (Provisional)*, London: War Office, 1920, pp. 14-15. J. F. C. Fuller was highly influential in the formulation of British doctrine and championed the principles of war. See J. F. C. Fuller, "The Principles of War, with Reference to the Campaigns of 1914-1915," *Journal of the Royal United Services Institution*, Vol. 61, February 1916, p. 18; J. F. C. Fuller, *The Foundations of the Science of War*, London: Hutchinson & Co. Publishers Ltd., 1926, pp. 226-227; and J. F. C. Fuller, *Armored Warfare*, Harrisburg: The Military Services Publishing Company, 1943, p. 11. A brief but informative comparative outline of multinational principles of war is found in Christopher Bellamy, *The Evolution of Modern Land Warfare: Theory and Practice*, New York: Routledge, 1990, pp. 13-14.

[3] For a synopsis of the evolution of unified command theater constructs in World War II, see James F. Schnabel, *History of the Joints Chiefs of Staff, Vol 1, The Joint Chiefs of Staff and National Policy 1945-1947*, Washington, DC: Office of Joint History, 1996, pp. 80-87.

[4] This reliance upon anecdotal evidence is regrettable but necessary to keep this paper unclassified. Evidence is drawn from interviews with U.S. and allied officers with experience in Afghanistan, HQ CENTCOM, HQ JFC Brunssum, HQ ISAF, and various other headquarters. The author's own experience is contributory in the part three analysis. He is a graduate of the U.S. Army C&GSC and SAMS, 2000; has worked within the USEUCOM Joint Planning Group 2001-2002; was a special adviser, plans to Commander ISAF V Lieutenant-General Hillier in 2004; commanded the Canadian Task Force Orion (OEF) in Kandahar 2006; was the Canadian Liaison Officer to HQ USCENTCOM 2006-07; and graduated from the U.S. Army War College in 2008.

[5] George W. Bush, *Unified Command Plan*, Washington, DC: The White House, May 5, 2006.

[6] W. E. Donaldson, *Military History Applied to Modern Warfare*, 2d Ed., rev. and enl. by A. F. Becke, London: Hugh Press, 1907, pp. 53-55: ". . . successful war is not waged by a congress but by one single commander."

[7] War Department, Office of the Chief of Staff, *Field Service Regulations*, U.S. Army, Washington, DC: U.S. Government Printing Office, 1914, pp. 67-68, in Article V, Combat Principles.

[8] Lieutenant Colonel T. M. Hunter, *Marshall Foch: A Study in Leadership*, Ottawa, Canada: Directorate of Military Training Army Headquarters, 1961, p. 186.

[9] For a review of the dynamics of supreme command under Foch, see Dominick Graham and Shelford Bidwell, *Coalitions Politicians & Generals: Some Aspects of Command in Two World Wars*, London: Brassey's, 1993, pp. 63, 142-149. Pershing objected strongly to Foch's demand for American divisional participation (under French command) in the offensives of September 1918. Pershing remembers countering this: "Marshal Foch, you may insist all you please, but I decline absolutely to agree to your plan. While our army will fight wherever you may decide, it will not fight except as an independent American army." Quoted in John J. Pershing, *My Experiences in the World War*, New York: Frederick A. Stokes Co., 2 Vols, 1931, Vol. II, pp. 246-247. General Foch recorded exchanges with Pershing in a more favorable light, remembering Pershing's offer of troops during the dark days of March 1918:

General Pershing had come to see me and, with magnificent spirit, spontaneously offered to throw immediately into the battle all the trained American divisions. General Bliss, the American representative on the Versailles committee, animated by a similar sentiment, exclaimed to me: "We have come over here to get ourselves killed; if you want to use us, what are you waiting for?"

Quoted in Ferdinand Foch, *The Memoirs of Marshal Foch*, Colonel T. Bentley Mott, trans., New York: Doubleday, Doran and Company, Incorporated, 1931, p. 270. Also see Hunter, *Marshall Foch*, pp. 195-196.

[10] Foch, *Memoirs*, p. xxvi.

[11] Pershing, Vol. I, p. 375.

[12] Russell F. Weigley, *History of the United States Army*, Bloomington: Indiana University Press, 1984, pp. 422-423. See also Graham and Bidwell, *Coalitions*, p. 159. For description of Foch's "strategic direction," see Foch, *Memoirs*, p. 275.

[13] The evolution of the principles of war in U.S. Army doctrine follows closely similar developments in Europe and is intricately connected to the progressive intellectual movement. The U.S. Army began in 1892 to attempt to codify enduring standards, establishing for the first time proper army regulations in War Department; *Troops in Company Regulations for the Army of the United States*, Washington, DC: U.S. Government Printing Office, 1892. These were augmented in the early 20th century by field regulations that addressed enduring standards in the conduct of war, beginning in 1905 with War Department; *The Field Service Regulations,*

United States Army, Washington, DC: U.S. Government Printing Office, 1905. The first discovered use of the principle of "unity of command" was found in the 1914 version of *Regulations*; see War Department, Office of the Chief of Staff, *Field Service Regulations, U.S. Army*, Washington, DC: U.S. Government Printing Office, 1914, pp. 67-68, in Article V Combat Principles. After World War I, the U.S. Army moved away from the field service regulations, replacing them with "training circulars," "training manuals," and "training regulations." In professional military curricula, the principles of war were officially adopted. However, the army was heavily influenced by European thinking, and the articulation of exact principles oscillated between American, British, and French percepts. J. F. C. Fuller's principles dominated in the immediate post-war era, and the American principle of "unity of command" was replaced in the 1920s by "cooperation." See War Department, *TR-10-5, Doctrine, Principles, and Methods*, Washington, DC: War Department, 1921, pp. 1-2. Fuller's influence however, only went so far in America. When British principles changed in 1923 from "cooperation" to "unity of effort," the U.S. Army retained "cooperation" until 1931. See Great Britain, War Office, *Field Service Regulations, Vol 1, Administration*, London; The War Office, 1923, pp. 4-6. Under Colonel George Marshall, Chief of Infantry in the 1930s, the U.S. Army reestablished "unity of command" as a key principle, first in infantry manuals. See U.S. Army Infantry School, *Tactics and Techniques of Infantry in Offensive Operations, United States Army Extension Course*, Fort Benning, GA: Publication Section of the Infantry School, 1931, pp. 2-4. The U.S. Army retained the principle of "cooperation" until 1939. See U.S. Army Command and General Staff School, *The Principles of Strategy for an Independent Corps or Army in a Theater of Operations*, Fort Leavenworth, KS: U.S. Army Command and General Staff School, 1936, p. 9. In 1939, Marshall's influence prevailed, and the Staff School replaced Fuller's "cooperation" with the French principle of "unity of effort" which stated: "Complete unity of a nation in war implies a single control over each effort and a uniting of all efforts under one head. In the armed forces, it is attained through unity of command." See U.S. Army Command and General Staff School, *The Offensive*, Fort Leavenworth, KS: U.S. Army Command and General Staff School, 1939, pp. 9-13. This was reiterated in the revamped Field Service Regulations of 1939. Here, "unity of effort" is defined as ". . . necessary to apply effectively the full combat power of the available forces. It is obtained through *unity of command*." See Chief of Staff of the U.S. Army, War Department, *Field Service Regulations (Tentative) Field Manual 100-5,* Washington, DC: U.S. Government Printing Office, 1939, pp. 27-29. This is changed again in the final version of the reinstituted regulations as the principle of "unity of command." See Chief of Staff of the U.S. Army, War Department, *Field Service Regulations Operations, Field Manual 100-5,* Washington, DC: U.S. Government Printing Office, 1941, pp. 22-23. The principle of "unity of command" has been accepted since this publication. In many doctrinal manuals the principle of "unit of effort" reappears periodically, not to replace "unity of command" but as a descriptive principle. For instance, see U.S. Army Command and General Staff College, *Field Service Regulations Operations, Large Units, ST 100-15-1*, Fort Leavenworth, KS: U.S. Army Command and General Staff College, 1963, p. 7. As well, the principle of unity of command has been somewhat diluted by the emergence of unified action doctrine in the 1990s, which too often overshadows "unity of command" in doctrine manuals. See Peter Pace, Chairman Joint Chiefs of Staff, *Joint Publication 3-0, Joint Operations,* Washington, DC: Joint Chiefs of Staff, September 17, 2006, p. xiii, GL-32. Here "unity of command" is forsaken in adherence to much more abstract "unified action." Only in descriptions of "unified command" does the principle of one supreme commander over joint forces shine through. However, despite muddying the doctrinal waters, the principle still remains the formal U.S. Army principle of war codified in U.S. military doctrine. See Department of the Army, *Operational Terms and Graphics, Field Manual 1-02 (FM 101-5-1) MCRP 5-12A,* Washington, DC: Headquarters Department of the Army, September 21, 2004, pp. 1-150.

[14] Maurice Matloff and Edwin Snell, *Strategic Planning for Coalition Warfare 1941-1942, United States Army in World War II*, Washington, DC: Office of the Chief of Military History, Department of the Army, 1953, pp. 123-124.

[15] Forrest C. Pogue, *George Marshall: Ordeal and Hope 1939-1942*, New York: The Viking Press, 1965, p. 375. See also Matloff and Snell, *Strategic Planning*, pp. 123-127. At the Arcadia Conference in December 1941, Marshall had been successful in convincing prominent opponents of the "unified command" concept (most notably Winston Churchill) of the necessity of putting Lieutenant-General Sir Archibald Wavell in charge of all multinational army, navy, and air force units in the newly designated Southwest Pacific Area, descriptively named the Australian-British-Dutch-American Command (or ABDA). This command was short-lived after the losses incurred in subsequent Japanese offensives, but the principle of unified commands endured.

[16] Marshall had witnessed the rivalry between Pershing and the CSA, General Peyton March, in 1917, due to Pershing's insistence that he worked directly for the President, through the Secretary of War, and not the CSA. His understanding of his command function was exactly that of General U. S. Grant—the only civil authority he need report to was the President. This was easy to argue in a war being fought by the United States on only one front. Marshall faced a war of multiple theaters of operations, each with competing demands for resources, and he understood the role of the JCS in defining which theater had priority. It is interesting to compare this understanding to the current UCP construct, where the Goldwater-Nichols Act has robbed JCS of the function

of prioritizing between theaters competing for requirements. For an interpretation of the Marshall-Pershing relationship, see Graham and Bidwell, *Coalitions*, pp. 159-163.

17 This system was an evolution from the Committee of Imperial Defence and the Imperial General Staff systems that had proved effective in mobilizing, training, equipping, and fighting millions of British imperial soldiers during World War I. See John Gooch, *The Plans of War: The General Staff and British Military Strategy c.1900-1916*, London: Routledge & Kegan Paul, 1974, pp. 131-160; and Graham and Bidwell, *Coalitions*, pp. 150-157.

18 Graham and Bidwell, *Coalitions*, pp. 150-156. See also Matloff and Snell, *Strategic Planning*, pp. 166-169.

19 Pogue, *George C. Marshall*, p. 380.

20 Matloff and Snell, *Strategic Planning*, pp. 156-173.

21 Forrest C. Pogue, *The Supreme Command, The United States Army in World War II*, Washington, DC: Office of the Chief of Military History, Department of the Army, 1954, p. 30.

22 *Ibid.*, pp. 48-52.

23 For a full treatment of this relationship, see James C. Humes, *Eisenhower and Churchill: The Partnership that Saved the World,* Roseville, CA: Prima Publishing, 2001.

24 USN support for geographic unified commands containing personnel from all services was predicated on reversing this decision in the Pacific. See Office of the Chairman of the Joint Chiefs of Staff, *The History of the Unified Command Plan 1946-1993*, Washington DC: Joint History Office, 1995, pp. 80-81.

25 Quoted from original UCP in *Ibid.*, p. 13.

26 *Ibid.*, pp. 15-16.

27 *Ibid.*, p. 12.

28 *Ibid.*, p. 2.

29 *Ibid.*, pp, 38-39. The panel also recommended "unfragmented command authority" by designating the component commanders as deputies to the unified commander to ensure that the chain of command was not being circumvented around the CINCs to the service chiefs.

30 Jack D. Nicholas, George B. Pickett, and William Spears, *The Joint and Combined Staff Officer's Manual,* Harrisburg, PA: The Stackpole Company, 1959, pp. 46-55.

31 Wesley K. Clark, *Waging Modern War*, New York: Perseus Books Group, 2000, p. 447. General Short, the Air Component Commander, was more frustrated when non-U.S. aircraft refused to bomb certain targets. As Clark states: "We paid a price in operational effectiveness by having to constrain the nature of the operation to fit the political and legal concerns of NATO member nations, . . . but the price brought significant strategic benefits." See p. 476. SACEUR himself was frustrated with alliance operations when his orders to seize the Pristina Airport (to prevent Russian entrance into Kosovo) were not executed by the British national commander, General Jackson (see pp. 394-403).

32 This was not the view of SACEUR, who is extremely clear in his commitment to the Alliance. See *Ibid.*, p. 447.

33 Robert H. Scales, *Certain Victory: The U.S. Army in the Gulf War*, Washington, DC: Office of the Chief of Staff of the U.S. Army, 1993, p. 122.

34 Secretary of Defense Donald Rumsfeld broke with American tradition and UCP practice by insisting upon redesignation of unified commanders from "commanders-in-chief" or "CINCs" to simply "commanders" in 2002. However, this paper will continue to use the term CINC in its original context to promote consistency of thought and practice and to avoid confusion.

35 These are a coalition force air component commander (CFACC), a coalition land force component commander (CFLCC), and a coalition force maritime component commander (CFMCC). For definition of Joint Force Commander (JFC) and breakdown of responsibilities of JFC and service component commanders, see Peter Pace, Chairman of the Joint Chiefs of Staff, *Joint Publication 1-0, Doctrine for the Armed Forces of the United States*, Washington, DC: Joint Chiefs of Staff, May 2006, pp. GL-8, and V-5 to V-18. For CLFCC, CFMCC and CFACC functions, see Peter Pace, Chairman of the Joint Chiefs of Staff, *Joint Publication 3-0, Joint Operations*, Washington, DC: Joint Chiefs of Staff, September 2007, p. GL 19-20.

36 Elaine M. Grossman, "Was Operation Anaconda Ill-fated From the Start?" *Inside the Pentagon*, Vol. 20, No. 31, July 29, 2004, reprinted in *Selected Readings, Academic Year 2008, Implementing National Security Strategy Vol. II*, Carlisle Barracks: U.S. Army War College, 2007, pp. 24-16 to 24–21. See also Sean Naylor, *Not a Good Day to Die: The Untold Story of Operation Anaconda*, New York: Berkley Book, 2005; and Major Mark G. Davis, "Operation Anaconda: Command and Confusion in Joint Warfare," unpublished thesis, Maxwell AFB, AL: School of Advanced Air and Space Studies, June 2004.

37 David W. Barno, "Fighting 'the Other War': Counterinsurgency Strategy in Afghanistan, 2003-2005," *Military Review*, September/October 2007, pp. 32-44.

38 The U.S. military perspective until late 2003 was that the politics of Kabul would interfere with proper military operations in Afghanistan and should be avoided. See *Ibid.*, p. 34.

39 Department of Implementing National Military Strategy, *Military Strategy, Planning, and Operations, Theater Organization and Command and Control Case Study: Operations Enduring Freedom and Iraqi Freedom, Supplemental Reading to Theater Organization and Command and Control*, Carlisle Barracks: U.S. Army War College, January 14, 2008, p. 11.

[40] David Barno, "Fighting," pp. 30-67.

[41] This is from personal experience of the author while participating in planning for ISAF expansion in 2004-2005 as a member of HQ ISAF V, working through JFC Brunssum. Commander JFC Brunssum refused continuously to acknowledge an operational level function inside of Afghanistan, believing that the operational level resided fully at Brunssum. This caused considerable friction between the two HQs, and further confused relations between these HQs and CFC-A.

[42] *Forbes.com*, "US to Reduce Troop Numbers in Afghanistan Soon," available from *www.forbes.com/finance/feeds/afx/2005/12/16/afx2414408.html*, Internet, accessed January 15, 2008. See also Associated Press and AFX News Limited, "NATO to Take Over Afghan Security in 2006," *USA Today*, August 4, 2005.

[43] U.S. constitutional and budgetary limitations, particularly Title 10 and Title 22 responsibilities held within specific CENTCOM service component headquarters, could not be mixed within NATO headquarters. Therefore, it was necessary to retain a parallel U.S. command structure, but this was done without any defining relationship with JFC Brunssum or SHAPE.

[44] CENTCOM commander still maintains responsibility for the following "Focus Areas" for Afghanistan: expand governance and security; support to counternarcotics efforts, resource and policy of PRTs; counterthreat finance work; and train and equip programs for Afghan security forces. Commander CENTCOM's authorities extend also to supporting efforts to degrade violent extremist networks in Afghanistan and Pakistan and the hunt for High Value Targets; roles assumed by SOCOM and STRATCOM under their own mandates. STRATCOM's functions in this fight include the provision of targeting support for—and prosecution of—global strike and global intelligence, surveillance and reconnaissance missions, controlled by JFCC Bolling Air Force Base. U.S. budgetary and constitutional obstacles prevent Commander CENTCOM from assuming overall control of counternarcotics and Afghan National Police training efforts (directed by DOS). Therefore, he works through his Joint Interagency Coordination Group (JIACG) for visibility onto these nonmilitary issues. There are no relationships between the JIACG and NATO. This situation is exacerbated by the fact that there has never existed a coherent interagency plan for Afghanistan, something that debilitates both U.S. interdepartmental cohesion and multinational efforts. See William Flavin, "Planning for Conflict Termination and Post-Conflict Success," *Parameters*, Vol. 33, No. 3, Autumn 2003, pp. 95-111.

[45] Pogue, *Supreme Command*, pp. 56-97.

[46] Commander Central Air Force (CENTAF – whose commander is also the CFACC supporting Afghanistan) has assumed responsibility for developing the Afghan Air Corps and created the Coalition Air Power Task Force (CAPTF) to execute this task. This Task Force is not part of CSTC-A, and has no relationship with NATO. CSTC-A also has no formal relationship with NATO, although its mission is essential to NATO tactical success.

[47] U.S. military authorities will state that allocation of Kandaks resides solely with the Afghan Minister of Defense; however, in practice CSTC-A and the Regional Command Advisory Groups control Kandak allocation, training, and operations cycles. NATO nations must lobby at the national level to gain Kandak partnerships.

[48] Throughout 2006, Canada, bearing most of the responsibility in RC-S, could not secure a single ANA company to partner with the large Canadian Task Force. As a result, that Task Force (Orion) conducted 27 offensive operations between January and August, supported by an average of only 20 ANA soldiers. Seven operations were conducted without any ANA involvement whatsoever, creating a situation where the only consistent "Afghan Face" on operations was that of the Taliban. A squad of senior Canadian generals had to broach this issue repeatedly with U.S. senior military authorities before Canada secured its first partnership, 15 months after deployment of forces to RC-S. The British experience in Helmand Province was the same. As the commander of Task Force Orion, the author can attest to the frustration and resentment this caused.

[49] This would also make it easier for commander ISAF to soften U.S. SOF resistance to creating partnerships between ANA units and NATO forces, U.S. SOF believing that NATO is ill-suited for Foreign Internal Defense (FID), and that NATO forces will withdraw within a few short years, leaving their Afghan charges unsupported. This attitude is reflected in that the majority of ANA Kandaks are still partnered with U.S. SOF units in Afghanistan. To NATO, the CSTC-A issue regarding allocations of Kandaks is indicative of a U.S. preference for a CENTCOM-run OEF mission over support for a NATO-run ISAF mission. On another issue, migrating CSTC-A to EUCOM would allow NATO members already contributing to the train and equip mission of CSTC-A to formally alleviate some of the 2 billion U.S. dollar price tag associated with building to ANA, while not jeopardizing U.S. accountability for Title 10 funding.

[50] At time of this writing, a military court of inquiry is still convened to look into the incident of March 4, 2007, in Nangahar Province where 19 civilians were killed and 50 injured during a Marine Corps special operations unit operation. The unit opened fire after its convoy was attacked by a suicide bomber. The Army brigade commander responsible for the region where the incident occurred, Colonel John Nicholson Jr., was not aware of the Marine special operations unit patrol in his area. "I was not aware that they were out doing a mission. We weren't sure who it was," said Nicholson. He later learned the unit had conducted more than 30 operations, only five of which he had been notified of at the time. Nicholson had to subsequently apologize to local Afghans for the deaths and approve payments to families. This incident reflects several that have occurred during the past 3 years involving poor coordination between special operations missions and

OEF/ISAF missions. See William M. Arkin, "Early Warning: Secret Operations: Supporting or Undermining the War on Terrorism?" available from *blog.washingtonpost.com/earlywarning/2008/01/secret_operations_undermine_th.html*; Internet, accessed January 15, 2008. See also "No Way to Win Hearts and Minds—America Apologizes for Killing Civilians," *Economist.com*, May 9, 2007; and Carlotta Gall and David E. Sanger, "Civilian Deaths Undermine Allies' War on Taliban," *New York Times*, May 13, 2007, p. 12.

[51] For instance, the 1942 revision of the 1936 U.S. Naval War College text cites the following:

> **Within the limits of human capacity, an organization can exert its combined effort with greater effect the more closely the exercise of command represents the act of a single competent commander. To divide the supreme command in any locality, or to vest it in a body rather than in an individual, is necessarily to diffuse responsibility. In that degree, there is then incurred the danger, through confusion of wills and ideas, of delaying decision and of creating corresponding diffusion of effort.**

See U.S. Naval War College, *Sound Military Decisions*, Annapolis, MD: Naval Institute Press, 1942, p. 12. See also Jack D. Nicholas, George B. Pickett, and William Spears, *The Joint and Combined Staff Officer's Manual*, Harrisburg, PA: The Stackpole Company, 1959, pp. 36-37, where is cited:

> **The first principle is that command must be vested in one individual. Applied to unified commands, this principle means that all organizational activities must conform to the fact that all assigned ground, naval, and air forces will be singly led and prepared to fight as one. The unified commander must have unquestioned authority over all units of his command to enable him to accomplish his mission and thereby discharge his responsibilities. Application of this principle further implies that the lines of authority be definite, clear-cut, and understood by all. Every individual should be indoctrinated with the fundamental concept of one boss at each level.**
>
> **There must be centralized direction and decentralized execution; no plural or divided authority over one area of responsibility; and a minimum of levels of authority, and diffusion of responsibility.**

[52] Quoted in Pogue, *Supreme Command*, p. 42. For discussion, see also Gary J. Borge, "Larger Units in North Africa and Mediterranean," in Gary L. Bounds, *Larger Units: Theater Army— Army Group—Field Army*, Fort Leavenworth, KS: Combat Studies Institute, 1985.

In: War in Afghanistan: Strategy…
Editor: Easton H. Ussery, pp. 53-121

ISBN: 978-1-60741-579-4

Chapter 3

War in Afghanistan: Strategy, Military Operations, and Issues for Congress

Catherine Dale
Specialist in International Security

Summary

As the 111th Congress convenes and the Obama Administration assumes office, the war in Afghanistan has entered its eighth year. With a deteriorating security situation and no comprehensive political outcome yet in sight, most observers view the war in Afghanistan as open-ended. By early 2009, a growing number of Members of Congress, Administration officials, and outside experts had concluded that the effort – often called "America's other war" – required greater national attention. For the Government of the Islamic Republic of Afghanistan (GIRoA), the war is both a struggle for survival and an effort to establish sustainable security and stability. For the United States, the war in Afghanistan concerns the security of Afghanistan and the region, including denying safe haven to terrorists and helping ensure a stable regional security balance. For regional states, including India and Russia as well as Afghanistan's neighbors Pakistan and Iran, the war may have a powerful impact on the future balance of power and influence in the region. For individual members of the North Atlantic Treaty Organization (NATO), the war may be about defeating terrorist networks, ensuring regional stability, proving themselves as contributing NATO members, and/or demonstrating NATO's relevance in the 21st century.

Since 2001, the character of the war in Afghanistan has evolved from a violent struggle against al Qaeda and its Taliban supporters to a multi-faceted counterinsurgency (COIN) effort. In the aftermath of the terrorist attacks of September 11, 2001, the United States launched Operation Enduring Freedom (OEF) in order to end the ability of the Taliban regime to provide safe haven to al Qaeda and to put a stop to al Qaeda's use of the territory of Afghanistan as a base of operations for terrorist activities. In that first phase, U.S. and coalition forces, working with Afghan opposition forces, quickly removed the Taliban regime.

After the fall of the Taliban, the character of the war shifted to a multifaceted COIN effort aimed at smothering the diffuse insurgency by shoring up GIRoA efforts to provide security, governance, and economic development. The three areas are generally viewed as interdependent and mutually-reinforcing – security is a prerequisite for some governance and development efforts, and longer-term, sustainable security requires both functional governance and economic opportunity. As one pillar of the COIN campaign in Afghanistan, the Afghan and international military effort aims broadly at defeating the remnants of the Taliban and other insurgents, securing the population, and helping extend the reach of the Afghan government. The international military effort includes both the NATO-led International Security Assistance Force (ISAF), to which the United States contributes troops, and the separate U.S.-led OEF mission.

The U.S. Government faces key strategic and operational decisions about its further engagement in the war in Afghanistan. These may include clarifying U.S. national interests in Afghanistan and the region; defining clear strategic objectives based on those interests; determining which diplomatic, economic, and military approaches to adopt, and what resources to commit to support those approaches; prioritizing "Afghanistan" versus other national security imperatives; and helping marshal a coordinated application of international efforts. Avenues available to Congress for exercising oversight include authorizing and appropriating funding for U.S. efforts in Afghanistan and the region; shaping policy through directive legislation; holding oversight hearings to assess policy execution; and adjusting Administration reporting requirements. This report provides analysis of current developments and future options concerning the war in Afghanistan. It will be updated as events warrant.

OVERVIEW

As the 111th Congress convenes and the Obama Administration assumes office, the war in Afghanistan has entered its eighth year. Unlike the war in Iraq, which, many argue, has entered its endstate, the war in Afghanistan – where the security situation has deteriorated and no comprehensive political outcome is yet in sight – appears to many observers to be open-ended. By early 2009, a growing number of Members of Congress, Administration officials, and outside experts had concluded that the effort – often called America's "other war" – required greater national attention.[1] In his inaugural address, President Obama stated that the United States would "forge a hard-earned peace in Afghanistan."[2]

For the Government of Afghanistan, the war is first of all an existential struggle for survival against the Taliban and other insurgents, as well as a longer-term effort to establish sustainable security and stability. For the U.S. Government, the war in Afghanistan concerns the security of both Afghanistan and the region, including denying safe haven to terrorists and helping ensure a constructive and stable regional security balance.[3] For regional states, including India and Russia as well as Afghanistan's immediate neighbors Pakistan and Iran, the war is critical because it may have a powerful impact on the future security and balance of power and influence in the region. For individual member states of the NATO Alliance, the war may be about some combination of defeating terrorist networks, ensuring regional stability, proving themselves as contributing NATO members, and demonstrating the relevance of the Alliance to 21st century security challenges.

Looking ahead, the U.S. Government faces major strategic and operational decisions about its further engagement in the war in Afghanistan. Those decisions may include:

- Clarifying U.S. national interests in Afghanistan and the region;
- Defining clear strategic objectives based on those interests;
- Determining which diplomatic, economic, and military approaches to adopt, and what resources to commit to support those approaches;
- Prioritizing the Afghanistan war versus other national security imperatives including the war in Iraq and preparing to meet potential threats; and
- Helping marshal a coordinated application of international efforts.

Avenues available to Congress for exercising oversight of these issues include authorizing and appropriating funding for U.S. efforts in Afghanistan and the region; shaping policy through directive legislation; confirming senior administration officials with responsibility for the Afghanistan effort; holding oversight hearings to assess policy formulation and execution; and extending or adjusting Administration reporting requirements.[4]

Character of the War in Afghanistan

While war is always about the organized use of violence to achieve political ends, the character of a given war may change dramatically over time. Since 2001, the character of the war in Afghanistan has evolved markedly, from a violent struggle against al Qaeda and its Taliban supporters, to a multi-faceted counterinsurgency (COIN) effort.

In the aftermath of the terrorist attacks of September 11, 2001, the United States launched Operation Enduring Freedom (OEF) in order to end the ability of the Taliban regime to provide safe haven to al Qaeda and to put a stop to al Qaeda's use of the territory of Afghanistan as a base of operations for terrorist activities. In that first phase, a primarily military effort, U.S. and other coalition forces, working closely with Afghan opposition forces, quickly removed the Taliban regime.

After the fall of the Taliban, the character of the war shifted to a multifaceted COIN effort aimed at smothering the diffuse insurgency by shoring up the efforts of the Government of the Islamic Republic of Afghanistan (GIRoA) to provide security, governance, and economic development. Leading practitioners view efforts in all three areas – and not just kinetic military operations – as essential to any successful counterinsurgency campaign. As U.S. Army General David Petraeus, now Commanding General of U.S. Central Command (CENTCOM), has frequently stated, "You can't kill your way out of an insurgency."[5] The three areas are generally viewed as interdependent and mutually-reinforcing – sufficient security is a prerequisite for some governance and development efforts, and longer-term, sustainable security requires both functional governance and economic opportunity. COIN theorists argue further that these areas require substantial civilian as well as military efforts. As the U.S. Army and Marine Corps 2006 COIN Manual states: "Military efforts are necessary and important to counterinsurgency efforts, but they are only effective when integrated into a comprehensive strategy employing all instruments of national power."[6]

As a central pillar of the COIN campaign in Afghanistan, the Afghan and international military effort aims broadly at defeating the remnants of the Taliban and other insurgents, securing the population, and helping extend the reach of the Afghan government. The international military effort now includes the North Atlantic Treaty Organization-led (NATO) International Security Assistance Force (ISAF), to which the U.S. Government contributes troops, as well as the separate US-led OEF mission.

Prospects for the Outcome of the War

Afghanistan's results to date have been mixed, and no concrete end of the war is yet in sight. Despite the achievement of some major political milestones – including ratifying a new constitution and holding presidential and parliamentary elections – progress to date in extending the rule of law, establishing effective governance, and furthering economic development has been relatively limited. Presidential elections, scheduled for 2009, are expected to further test the fabric of society. Meanwhile, for several years, practitioners and observers have expressed concerns about a worsening security situation on the ground, including the greater frequency and sophistication of attacks, exacerbated by the ability of insurgents to find safe haven across the border in Pakistan.

Experts differ on the further prospects for the Afghanistan effort and the war's likely outcome, in part because they pose the question in different ways. One approach addresses the relatively short-term goal of defeating the insurgency – that is, ensuring that insurgents cannot directly challenge the authority of the Afghan state.

As of late 2008, few if any practitioners or observers expected the war to end in a clear Taliban victory, including Taliban control of the state of Afghanistan. Some suggested that a more likely worst-case scenario would be a reversion to the civil war and chaos of the early 1990's, including warlordism, a general lack of stability and opportunity for ordinary Afghans, and a proliferation of ungoverned spaces that might be used by terrorists as safe havens. To some extent, these conditions are currently manifested in parts of southern Afghanistan.

In late 2008, as a rule, U.S. and other international senior officials in Afghanistan expressed measured optimism regarding near-term results of the counterinsurgency effort. They pointed to some recent progress breaking down insurgent networks and expected further gains, particularly if more resources were made available and greater cooperation from all parties, including neighboring states, achieved. As a rule, international officials did not argue that without more resources, the COIN effort would fail, but rather, that without more resources, the effort would cost more money, more time, and more lives.[7] Some members of this school of thought offered the caveat that making timely progress was important, because a lingering stalemate with the insurgency would be likely to spur greater popular disaffection.

Another approach to the question of Afghanistan's prospects takes a broader and longer-term view. Observers from this school of thought point to thirty years of war, occupation, displacement and chaos that have destroyed Afghanistan's infrastructure, ravaged its human capital resources, and left most of its relatively youthful population with no memories of living in a society not disrupted by conflict. Some experts caution that even if the insurgency is defeated in the near-term, it is not hard to imagine that some remnants or some later generations might draw inspiration from the current fight, and resume the attack whenever the political and security constellation seems more conducive to their success – as Afghans say, "You have the watches; we have the time." Other observers note that in comparative and historical global perspective, it is quite rare for states to achieve "stability" and "good governance", and that Afghanistan, given its relative poverty of human and natural resources, faces steep challenges and unlikely odds in aiming at those objectives. Accordingly, some senior U.S. and other international officials have urged a tempering of expectations about Afghanistan's long-term prospects.

Purpose of This Report

This report provides:

- A concise summary of the war's background, context, and early execution;
- An analytical discussion of the COIN war to date, including strategy, organization, participation, and key facets of the effort including population security, advising the Afghan National Security Forces (ANSF), counter-narcotics, reconciliation, community outreach, and civil-military coordination; and
- An analysis of major strategic and operational issues and options that the 111th Congress may wish to consider.

Background: Context and Early History of the War

Current efforts to support security, governance and development in Afghanistan take place in the aftermath of thirty grueling years of conflict and unrest, followed by OEF military operations that removed the Taliban regime and the rapid creation of a new, post-Taliban political order.

Prelude to War[8]

In December 1979, the Soviet Union invaded Afghanistan to shore up a puppet communist regime. During the 1980's, armed Afghan resistance groups known as "mujahedin" waged war against Soviet forces and the Afghan security forces that supported them.[9] During that period, the U.S. Government, through the Central Intelligence Agency (CIA), provided covert assistance to mujahedin groups, working through Pakistan's Inter-Services Intelligence Directorate (ISI).

In 1989, Soviet forces withdrew from Afghanistan, and in April 1992, the Soviet-backed Afghan regime in Kabul fell to mujahedin forces, which established a form of rule including a rotating presidency. In November 1994, the ethnically Pashtun-dominated Taliban movement led by Mullah Omar seized the city of Kandahar, in southern Afghanistan.[10] In 1996, the Taliban captured Kabul and retained control over much of the country until ousted by OEF operations in 2001. However, throughout its tenure, the Taliban continued to face armed opposition, in particular from the Northern Alliance, a loose network dominated by ethnic Tajiks and Uzbeks, primarily from northern Afghanistan. Key legacies of Afghanistan's years of civil war, conflict, and oppressive rule included the deaths of over a million people, the displacement of millions more, the proliferation of available weapons, and the destruction of key institutions and infrastructure.

The proximate cause of U.S. military operations in Afghanistan was the linkage of the September 11, 2001, terrorist attacks to al Qaeda, which trained and operated under Taliban protection in Afghanistan. In an address to a joint session of Congress on September 20, 2001, President George W. Bush stated U.S. demands on the Taliban, warning: "The Taliban must act, and act immediately. They will hand over the terrorists or they will share in their fate."[11]

Major Combat Operations

On October 7, 2001, following the refusal of the Taliban regime to cease harboring al Qaeda, the U.S. Government launched military operations in Afghanistan, with the stated purpose of disrupting the use of Afghanistan as a terrorist base of operations and attacking the military capability of the Taliban regime.[12]

In contrast to the lengthy, iterative preparations that preceded the launch of Operation Iraqi Freedom, the U.S. planning process for OEF was extremely condensed. The concept of operations was based on Secretary of Defense Donald Rumsfeld's vision of defense transformation, including the idea that a heavier reliance on cutting-edge technology and precision weaponry could make possible the deployment of smaller-sized conventional ground forces.

Military operations were preceded and complemented by work by the Central Intelligence Agency (CIA) with Afghan opposition groups on the ground. Initial U.S. operations relied on the use of special operations forces (SOF) on the ground, enabled by air assets, working by, with and through indigenous partners, in particular the Northern Alliance. Many U.S. defense experts regarded the operations as an important demonstration of operational "jointness" – the ability of Military Services to work together seamlessly. The United Kingdom and Australia also deployed forces to support the major combat phase of operations, and dozens of other countries provided basing, access and overflight permission.[13]

Military victory, including the demise of the Taliban regime, came quickly. In November 2001, the Taliban fled Kabul, and in December they left their stronghold, the southern city of Kandahar. It is generally understood that in December 2001, key al Qaeda and Taliban leaders fled across the border into Pakistan.

To fill the political void, in December 2001, in Bonn, Germany, the United Nations hosted the so-called Bonn Conference. Participants included representatives of four Afghan opposition groupings, and observers included representatives of neighboring and other key countries including the United States. The resulting Bonn Agreement created an Afghan Interim Authority to serve as the "repository of Afghan sovereignty" and outlined a political process for producing a new constitution and choosing a new Afghan government. In contrast to the model pursued in Iraq from 2003 to 2004, in Afghanistan there was no period of formal occupation in which an international authority exercised sovereignty on behalf of the Afghans.[14] To help provide security to support the fledgling new regime, in December 2001 the United Nations authorized an international force – the International Security Assistance Force (ISAF) – with a mandate to help the Afghans maintain security in Kabul and surrounding areas. The United Kingdom agreed to lead the force initially.[15]

The major combat operations phase was regarded as a quick success by its Afghan protagonists and their U.S. and other international partners, but the challenges were far from over. The new Afghan leadership faced the profound political challenge of consolidating a fractious, scarred state, with very few resources. The new leaders also faced potential violent challenges, both from resurgent al Qaeda and Taliban leaders who were defeated but not eliminated, and from Afghan local power-brokers, strengthened by years of battle-hardened autonomy and resistance, who were displeased by the emerging post-Taliban order.

Counterinsurgency War in Afghanistan to Date

Both the security climate – including the composition, strategy, and tactics of the insurgency – and the structure and focus of international efforts designed to support the Afghan government have changed substantially since the end of the major combat operations that ousted the Taliban regime. This section describes and analyzes key developments in the counterinsurgency war in Afghanistan with an emphasis on recent trends and initiatives.

Strategy

In 2008, as international interest in, and attention to, the war in Afghanistan grew, a number of observers stressed the need for clearer or more robust strategy to guide Afghan and international efforts. "Strategy" is commonly understood to include a statement of objectives, or desired ends; the ways and means designed to achieve those ends; and the roles and responsibilities of key players in executing those ways and means.[16]

Strategy-making for Afghanistan is particularly complicated, for two main reasons. First, the range of strategic objectives is quite broad, encompassing not only security progress but also, for example, civilian capacity-building, the rule of law, counternarcotics, and economic development. Those fields, in turn, are closely linked empirically – for example, long-term development requires a relatively stable environment, and successful counternarcotics efforts must be predicated on some form of rule of law. Second, strategy-making is complicated by the range of actors providing some support to GIRoA, including NATO, the United Nations, and other international organizations, as well as individual states, each of which may have its own – or even competing sets of – interests and priorities. Military strategy, in turn, is not easily separable from broader grand strategy for Afghanistan, since security is essential for progress in other areas, and since military forces play key supporting roles in the non-security lines of operation.

NATO Strategy

At its 20th Summit, held in Bucharest, Romania, in April 2008, NATO issued a streamlined but clear strategic vision for Afghanistan. That vision established four "guiding principles": a firm and shared long-term commitment; support for enhanced Afghan leadership and responsibility; a comprehensive approach by the international community, bringing together civilian and military efforts; and increased cooperation and engagement with Afghanistan's neighbors, especially Pakistan. The document also included a "vision of success", which is essentially a statement of objectives: "extremism and terrorism will no longer pose a threat to stability; Afghan National Security Forces will be in the lead and self-sufficient; and the Afghan Government will be able to extend the reach of good governance, reconstruction, and development throughout the country to the benefit of all its citizens."[17] What the "strategic vision" did not provide in any detail was a clear articulation of the specific ways and means ISAF would use to achieve those objectives.

Arguably closing the strategy gap substantially, ISAF, in October 2008, issued a classified Joint Campaign Plan (JCP). The JCP was GEN McKiernan's guidance to the force, and it specified key assumptions, objectives, and approaches to be used to achieve those objectives. It stated that the primary goal is the "transfer of lead security responsibility" to the Afghans, which includes planning as well as conducting operations. The JCP addressed all the lines of operation (LOO) discussed in the Afghanistan Compact but underscored that NATO has the lead only for the security LOO. Importantly, the JCP framed ISAF's mission in counterinsurgency (COIN) terms – the mission includes defeating an "insurgency" and the basic approach follows the COIN logic of "shape, clear, hold, build." ISAF officials considered the use of COIN terminology a breakthrough, following years of NATO preference for framing the effort in Afghanistan in terms of stability operations.[18]

U.S. Government Strategy

The U.S. Government plays a significant leadership role in both ISAF and NATO as a whole, and thus helps shape NATO and ISAF strategy and approaches. At the same time, the United States may have national interests in Afghanistan and the region that are not shared by all ISAF contributors, and the relative priority of various interests may differ among the Allies.

As of the start of 2009, the U.S. Government had not yet published a formal strategy for Afghanistan, along the lines of the November 2005 *National Strategy for Victory in Iraq*.[19] However, in the context of January 2009 confirmation hearings, key Obama Administration officials gave indications about the likely content of future U.S. strategy. Nominated to the post of Under Secretary of Defense for Policy, Michèle Flournoy stated that "our strategic objective is a stable and secure Afghanistan in which al Qaeda and the network of insurgent groups, including the Taliban, are incapable of seriously threatening the Afghan state and resurrecting a safe haven for terrorism."[20] Secretary of State Hillary Clinton indicated that President Obama's Afghanistan strategy was likely to include these elements: sending additional troops to Afghanistan; providing a "major increase" in non-military aid to Afghanistan; confronting the drug trade; and developing a coherent Pakistan policy. Furthermore, based on a policy of "more for more," aid to GIRoA would be tied to better performance.[21]

Meanwhile, CENTCOM has been conducting a 100-day comprehensive strategic review of its entire area of responsibility, including Afghanistan. For his part, GEN McKiernan, noting that the views were his own, has suggested that U.S. interests might include ensuring that Afghanistan cannot harbor terrorists; establishing a controlled Afghanistan/ Pakistan border; promoting a degree of regional stability; supporting a constructive role for Iran; and encouraging some form of freedom and democracy for the Afghan people.[22] These ideas and initiatives may contribute to a new articulation of U.S. strategic objectives regarding Afghanistan early in the Obama Administration.

International Efforts: Organization and Coordination

Afghanistan, which lacks sufficient institutional, material and human resources to make substantial progress on its own, relies deeply on the international community to support the three main pillars of the counterinsurgency effort: security, governance and development. However, most practitioners and observers contend that ever since the Bonn Conference, the multi-faceted international effort has suffered from a dearth of resources in each area, and from insufficient coordination among key players and their initiatives.

The "lead nation" model of international assistance to Afghanistan was agreed to at a donors' conference held in Tokyo in early 2002. Five countries each agreed to assume lead coordination responsibility for assistance to a single area of security-related Afghan administration: the United States for the army, Germany for the police, Italy for the judiciary, the United Kingdom for counternarcotics, and Japan for the disarmament, demobilization and reintegration (DDR) of militias.

The Afghanistan Compact, a formal statement of commitment by the Government of the Islamic Republic of Afghanistan (GIRoA) and the international community, finalized in January 2006, shifted responsibility from lead nations to Afghanistan itself. The premise was a shared Afghan and international vision of Afghanistan's future, including the commitment of the international community to "provide resources and support" to realize that vision. The Compact established three broad pillars of activity for future efforts – security; governance, the rule of law and human rights; and economic and social development. To "ensure overall strategic coordination of the implementation of the Compact," the document established the Joint Coordination and Monitoring Board (JCMB), co-chaired by a GIRoA representative and the United Nations Special Representative of the Secretary-General (UN SRSG).[23]

The UN SRSG leads the United Nations Assistance Mission in Afghanistan (UNAMA), which was established by the UN Security Council in early 2002.[24] The current UNAMA mandate confirms the UN SRSG's lead coordination role, as described by the Afghanistan Compact, but clarifies that the UN plays a stronger coordination role vis-à-vis civilian assistance efforts, than for military ones. The mandate states that the UN SRSG will "lead the international civilian efforts" to promote "...more coherent support by the international community to the Afghan Government." Concerning military efforts, the UN SRSG will work to "strengthen cooperation with ISAF at all levels."[25]

According to UN SRSG Kai Eide, one challenge the UN faces in performing its coordination function in Afghanistan is that UNAMA is under-resourced and under-staffed to play that role well.[26] An additional and more subtle challenge may be an inherent tension between, on one hand, UNAMA's mandate to coordinate international assistance efforts and help ensure that the international community speaks to GIRoA with a unified voice; and on the other hand, the UN SRSG's role as an advocate for human rights, in which capacity the UN mission may need to preserve a measure of independence from, and the option of critiquing, international military activities. In late 2008, UN SRSG Eide stressed the need to preserve the perception of UNAMA's role as an "honest broker."[27]

Security Line of Operation: Organization

International military forces in Afghanistan lead support to GIRoA in the field of security – one of the three pillars of the Afghanistan Compact – and support international civilian initiatives in the other two fields, governance and development. Over time, the mandates, structure and composition of the international force presence in Afghanistan have changed significantly, as the role of NATO has increased and the character of the fight has evolved. Today, NATO leads the International Security Assistance Force (ISAF), and the United States leads the OEF coalition effort. The U.S. government contributes troops to both missions.

NATO International Security Assistance Force (ISAF)[28]

ISAF represents NATO's first significant out-of-area deployment, and it is viewed by many observers as a key test for the Alliance – a measure of both its current capabilities and its possible future relevance. On September 12, 2001, in response to the 9/11 terrorist attacks, NATO for the first time invoked Article 5 of the North Atlantic Treaty, which confirms the commitment of the allies to collective self-defense in the event of armed attack on any party to the treaty.[29] That action helped clear the way for future NATO operations in Afghanistan. On August 9, 2003, NATO assumed responsibility for the ISAF mission, which had been established by UN mandate in December 2001 and led until mid-2003 by a series of lead nations.

ISAF Stages

ISAF, initially mandated to support Afghan efforts to secure Kabul and its immediate environs, expanded its geographical scope in four stages. During Stage 1, completed on October 1, 2004, ISAF expanded to the north of Kabul, assuming responsibility for a German-led Provincial Reconstruction Team (PRT) and establishing new PRTs. In Stage 2, completed in September 2005, ISAF expanded to the west. In Stage 3, completed on July 31, 2006, ISAF assumed responsibility for southern Afghanistan. In Stage 4, completed on October 5, 2006, ISAF assumed control of U.S.-led forces in eastern Afghanistan.[30]

ISAF Mandate

The focus of ISAF efforts has been a source of contention among the Allies, many of whom agreed to contribute troops on the premise that ISAF's focus would be post-conflict stability operations. That premise may have been valid at the time of ISAF's formation, but by several years later, the security climate had changed and an organized, capable insurgency had emerged. ISAF Commanding General, U.S. Army General David McKiernan, stated in November 2008: "The fact is that we are at war in Afghanistan. It's not peacekeeping. It's not stability operations. It's not humanitarian assistance. It's war."[31] ISAF's mission statement reflects the insurgency challenge: "ISAF conducts operations in partnership with GIRoA and in coordination with OEF, UNAMA, and the international community in order to assist GIRoA to defeat the insurgency, establish a secure environment, extend viable governance, and promote development throughout Afghanistan."[32]

ISAF Phases

From the outset, NATO planned that ISAF operations in Afghanistan would have four phases. The first phase was "assessment and preparation", including initial operations only in Kabul. The second phase was ISAF's geographic expansion throughout Afghanistan, completed in 2006. The final three phases would be stabilization; transition; and redeployment. At the start of 2009, ISAF was operating in Phase III, "stabilization", and NATO officials were reportedly discussing when to announce the commencement of Phase IV, the "transition" of lead security responsibility to the Afghan National Security Forces (ANSF). Some ISAF officials have expressed the concern that an announcement that ISAF has entered "transition" could trigger a rush by some troop-contributing countries to Phase V – "redeployment". They caution that in practice, the shift from stabilization to transition is likely to vary geographically across Afghanistan as the abilities of various ANSF to execute and then lead missions increase, and to take place in fits and starts, rather than at a clear single point in time.[33]

ISAF Organization

ISAF is led by a four-star combined headquarters, based in Kabul and headed by U.S. Army General David McKiernan. NATO's North Atlantic Council provides political direction for the mission. NATO's Supreme Headquarters Allied Powers in Europe (SHAPE), based in Mons, Belgium, and led by Supreme Allied Commander Europe (SACEUR), U.S. Army General John Craddock, provides strategic command and control. NATO's Joint Force Command Headquarters, which is based in Brunssum, The Netherlands, and reports to SHAPE, provides "overall operational control", including many administrative responsibilities. ISAF itself, which reports to SHAPE through Joint Forces Command, exercises "in-theater operational command". This arrangement, including two levels of operational headquarters, is somewhat unusual.

In Afghanistan, ISAF oversees five contiguous Regional Commands (RC), each led by a two-star general: RC-Center, led by France; RC-North, led by Germany; RC-West, led by Italy; RC-South, under rotating lead by Canada, the Netherlands, and the United Kingdom; and RC-East, led by the United States. Troop contingents from other Allies, and from some non-NATO partners, serve under these Regional Commands.

ISAF Troop Contributions

As of January 12, 2008, ISAF included 55,100 troops from 41 countries, including Allies and non-NATO partners.[34]

From the outset, NATO has struggled to secure sufficient troop contributions for ISAF. One consideration for potential troop contributors is cost – NATO's long-standing practice, "costs lie where they fall", typically means that countries pay their own costs when they contribute troops to a mission such as Afghanistan. Another consideration is the need for domestic political support. For some countries, this may require the ability to point to progress on the ground in Afghanistan to prove that the effort can yield results. As a senior Allied officer in ISAF explained, "right now it's all bad news about Afghanistan, so we need to show that something is possible."[35]

In 2008, and particularly during preparations for the April NATO Summit held in Bucharest, Romania, the U.S. Government actively sought pledges of additional troop contributions. While the Summit yielded pledges from Allies and partners for over 2,000 non-U.S. troops, not all of those pledges have been realized, and the pledge total fell short of the requests by ISAF commanders on the ground.[36] In December 2008, the United Kingdom announced the deployment– already accomplished – until August 2009, of an additional 300 special operations forces to serve in restive Helmand province in southern Afghanistan.[37] At the start of 2009, the UK Government was reportedly considering the temporary deployment to Afghanistan of 3,000 additional British troops.[38] ISAF's planning assumption, in early 2009, was that NATO troop contributions were likely to remain steady for the next several years.

ISAF National Caveats

From the outset, ISAF operations have been constrained by "national caveats" – restrictions that individual troop-contributing countries impose on their own forces' activities. Caveats tend to be informed by domestic political constraints – a government may consider, for example, that only by limiting its troops' activities, and hedging against taking casualties, can it guard against strong popular domestic opposition to its troop contribution. As a rule, as a rule, troop-contributing countries state their caveats explicitly; but additional constraints may surface when unanticipated requirements arise and contingents seek additional guidance from their capitals.

The nature and extent of national caveats varies greatly among ISAF participants. Senior U.S. military officials point with concern, for example, to constraints on German forces in Afghanistan, which are imposed by Germany's parliament the Bundestag. These include restrictions on German training and advisory teams that do not allow them to conduct combined offensive operations with their Afghan counterparts, and on capable German Special Operations Forces (SOF) that are "FOB-locked", that is, effectively confined to their Forward Operating Base. Not all contingents are so constrained. U.S. officials praise, for example, the 700-strong French infantry battalion that works closely with U.S. SOF and Afghan counterparts in Kapisa province, at the "north gate" into Kabul, which witnessed growing insurgent infiltration in 2008.

National caveats frustrate commanders on the ground because they inhibit commanders' freedom to apportion forces across the battlespace – to move and utilize forces freely. With caveats, the "whole" of the international force, as some observers have suggested, is less than the sum of its parts. Even more damaging, ISAF officials note, is the impact caveats can have on ISAF's relationship with Afghan National Security Forces (ANSF) counterparts. For example, ISAF advisory teams that are unable to accompany ANSF counterparts on offensive operations quickly lose both the Afghans' respect, and their own ability to shape and mentor the Afghan forces. Afghan Minister of Defense Abdul Rahim Wardak stated that ISAF training teams "don't have the same quality" as their U.S. counterparts.[39] U.S. senior military officials in Afghanistan frequently note that the ANSF appreciate their U.S. counterparts because "we drink from the same canteen."[40]

At the 20th NATO Summit, held in Bucharest, Romania, April 2-4, 2008, the Bush Administration strengthened its efforts to urge ISAF troop contributors to drop or ease their national caveats, with some success.

Coordination Within NATO/ISAF

ISAF officials note that both command and control, and coordination, within the NATO mission in Afghanistan leave some room for improvement.

One challenge is ensuring a full command relationship between ISAF headquarters and the Regional Commands. In RC-South, for example, the major troop contributors – the UK, Canada, the Netherlands – are strong partners relatively unconstrained by caveats. But ISAF officials note that RC-South effectively includes four provincially-based national campaigns – Dutch, British, Canadian, and U.S. – based on the provinces in which their respective troops are deployed. Each of these ISAF countries, in turn, tends to lobby the relevant Afghan Ministers in Kabul for assistance to "its" province. The RC-South commander, ISAF officials underscore, has never been empowered to give comprehensive guidance to the other nations in that RC command. In November 2008, the new RC-South Commanding General, Major General de Kruif from the Netherlands, indicated the relatively low standard of expectations, stating: "it's not about unity of command, but about unity of effort."[41] GEN McKiernan has stressed, however, that a more closely integrated effort is necessary, not least because insurgents and tribes do not define their efforts by provincial or district boundaries.[42]

An additional challenge is information flow among ISAF participants. Senior U.S. officials at the ISAF HQ in Kabul note that they have a much clearer operational picture of eastern and southern Afghanistan, where most U.S. forces operate, than of northern and western Afghanistan. Constraints on information flow may include the use of different – national and NATO – communications channels, linguistic barriers, and some reluctance on the part of some countries to share information perceived to be especially sensitive.

Another major challenge is maintaining situational awareness of – let alone control over – the activities of the 26 nationally-sponsored Provincial Reconstruction Teams (PRTs), which foster the ability of Afghan provincial-level officials to provide and promote governance, development and security. Officially, the military component of each PRT falls under ISAF command.[43] In practice, despite an ISAF directive that the PRT military components must report to ISAF on their activities, information flow, as of late 2008, was spotty. One senior ISAF official speculated, in late 2008, that one reason for the historical failure to comply might have been the perception that such efforts were wasted – that is, that the ISAF HQ made little use of such data and provided nothing back to the PRTs in return. The lack of a clear, shared picture of PRT activities has frustrated not only the ISAF leadership, but also Afghan and UN officials, in their efforts to apply resources strategically and effectively.[44]

In November 2008, the ISAF HQ restructured PRT oversight, to include the capability to share lessons learned and provide analytical feedback to PRTs. And in December 2008, a long-moribund Executive Steering Committee, including senior leadership from ISAF, GIRoA, and UNAMA, was reconstituted. As a result, some officials are optimistic that information flow concerning PRTs may improve in 2009.

One further challenge to full ISAF unity of command is the distinct mandate and role of the NATO Senior Civilian Representative (SCR), a position held since July 2008 by Italian Ambassador Fernando Gentilini. The SCR is the representative in Afghanistan of the NATO Secretary-General and reports regularly to the North Atlantic Council. As described by Ambassador Gentilini, the purpose is to "show that NATO is not just a military organization but that it can contribute to the political process more broadly."[45] Some ISAF senior officials view the SCR position – not necessarily any specific incumbent – as a "free agent", since the

SCR is not part of ISAF, and ISAF and the SCR are not required to speak to counterparts in Afghanistan with a single NATO voice.[46]

U.S. Forces in Afghanistan

The U.S. footprint in Afghanistan, and command and control arrangements for U.S. forces deployed there, have evolved over time, partly in response to the expansion of ISAF's area of responsibility to include all of Afghanistan.

U.S. Command Structure

Since major combat operations in 2001, the U.S. military has maintained a distinct special operations forces (SOF) presence in Afghanistan, reporting to U.S. Special Operations Command (SOCOM). By early 2002, some U.S. conventional forces, including a two-star U.S. Army Division Headquarters, had flowed into Afghanistan, but the footprint remained light – only one brigade combat team (BCT) – until early 2007.

In October 2003, the U.S.-led three-star Combined Forces Command-Afghanistan (CFC-A) was established in Kabul. CFC-A oversaw two U.S.-led two-star commands that also included coalition partners – a training command for the ANSF, and a Combined Joint Task Force (CJTF) of conventional forces in eastern Afghanistan. CFC-A served until ISAF assumed security responsibility for all of Afghanistan, and was then deactivated, in February 2007. Following the deactivation of CFC-A, its subordinate ANSF training command, the Combined Security Transition Command-Afghanistan (CSTC-A), began reporting directly to U.S. Central Command (CENTCOM), and its subordinate CJTF assumed a dual U.S./NATO reporting chain, to CENTCOM for U.S. issues and to ISAF in its NATO capacity as RC-East. In October 2008, the Department of Defense activated United States Forces-Afghanistan (USFOR-A), a new four-star headquarters designed to streamline command and control for U.S. forces operating in Afghanistan. ISAF Commanding General GEN McKiernan was given the additional assignment of serving as the USFOR-A Commanding General. As the head of ISAF, GEN McKiernan reports up the NATO chain of command to SACEUR GEN John Craddock; as the head of USFOR-A, he reports to the Commanding General of CENTCOM, GEN David Petraeus.

U.S. Force Levels

As of December 10, 2008, the United States had 36,234 total forces deployed in Afghanistan, including 22,658 in ISAF, and 13,576 in USFOR-A. The USFOR-A forces included CSTC-A. The ISAF forces included the 101st Airborne Division Headquarters, serving as the RC-East HQ, plus two BCTs and an aviation brigade.

The December 2008 numbers marked a significant increase from two years earlier, in December 2006, when U.S. forces in Afghanistan included only one BCT. In early 2007, an additional BCT was added, by extending the tour of the 3rd BCT, 10th Mountain Division (3/10) by 120 days, flowing in its original replacement, 4th BCT, 82nd Airborne Division, on schedule, and later replacing 3/10 with the 173rd Airborne BCT.[47] In January 2008, the Department of Defense announced that President Bush had approved an "extraordinary, one-time" deployment of 3,200 additional Marines to Afghanistan.[48] Those forces included the 24th Marine Expeditionary Unit (MEU), which served as a combat force in southern Afghanistan, and the 2nd Battalion, 7th Marine Regiment (2/7) who served as advisors for the

ANSF. Both units redeployed in November 2008, but a Marine Air Ground Task Force (MAGTF), including 3rd Battalion, 8th Marine Regiment, plus additional logistics and air support, deployed to southern Afghanistan in November 2008 to serve both as battlespace owners – responsible for security in a given area of operations – and as ANSF advisors.

In 2008, GEN McKiernan requested significant additional forces to serve in RC-East and RC-South. The first result was the deployment of the 3rd BCT, 10th Mountain Division, in January 2009. That brigade is serving in Wardak and Logar provinces, marking the southern entrances to Kabul, which witnessed growing insurgent infiltration and activity in 2008.

For RC-South, GEN McKiernan requested a U.S. force package similar to the one in RC-East – that is, three BCTs (or equivalents), an aviation brigade, and key enablers including engineers. He has argued that in southern Afghanistan, sufficient international and Afghan security forces are simply not available to "provide for adequate security for the people."[49] U.S. military officials in Afghanistan noted in late 2008 that some areas of southern Afghanistan contained known security challenges that still needed to be addressed – these included Kandahar city, and a part of Garmsir District in Helmand province known as the "fish hook" and long used by insurgents as a base of operations. Other areas, including Nimroz province, were "unknowns" given the lack of international and Afghan forces deployed there. Further areas simply had too few forces to fully clear and hold them – for example, when ISAF assumed responsibility for southern Afghanistan, Romanian forces replaced U.S. forces in Zabul province, but the Romanians had been prepared to conduct stability, not counterinsurgency, operations.[50]

The Department of Defense approved the request for some additional forces, pending their availability. On a visit to Afghanistan in December 2008, Secretary of Defense Robert Gates stated that the Department of Defense (DoD) hopes to send an additional two BCTs by late spring 2009.[51] Those additional forces, according to U.S. commanders in Afghanistan, are expected to flow primarily into Kandahar, Helmand, and Zabul provinces.[52]

U.S. Leadership

In addition to the need for additional U.S. forces, GEN McKiernan and other U.S. commanders have stressed the need for additional U.S. leadership in southern Afghanistan, to help synchronize and strengthen ISAF efforts. "Without U.S. leadership," GEN McKiernan stated in late 2008, "we will not break out of the stalemate in the south."[53] In November 2008, a new billet – a Deputy Commanding General for Stability – was created at RC South and filled by a U.S. Army Brigadier General. The United States is reportedly scheduled to take command of RC-South in 2010.[54]

Key Enablers

U.S. commanders and officials in Afghanistan stress the need for sufficient enablers to support the growing force. Engineers are critical, they underscore, to support both the construction of any additional defense infrastructure required by the deployment of additional forces, and to play a supporting role in reconstruction efforts.

Aviation is critical for both combat operations and also – especially – for air mobility in a country whose lack of infrastructure and forbidding terrain severely limit the utility of ground transportation. Helicopters, GEN McKiernan has stated, are ISAF's "trucks, our horses", for moving around the battlespace.[55] Ground vehicles, in turn, must be well-suited for their proposed use on Afghanistan's rugged terrain. Some U.S. troops have reportedly found that

Mine-Resistant Ambush-Protected (MRAP) armored fighting vehicles, which have provided life-saving protection against improvised explosive devices (IEDs) in Iraq, are less well-suited to Afghanistan's unpaved roads and off-road requirements.[56]

Current Intelligence, Surveillance and Reconnaissance (ISR) assets are, according to one U.S. commander, "not even in the ballpark", and, according to a senior ISAF official, ground units "are screaming for more assets."[57] A former U.S. battalion commander in Afghanistan argued after his tour, "As a rule, each battalion-sized task force should have constant unmanned aerial vehicle and close-air-support coverage."[58] In 2008, Secretary Gates, recognizing a need to provide troops in the field with improved ISR assets, formed an ISR task force to assess requirements and speed the process of meeting warfighter needs.[59] A CENTCOM ISR Task Force is in the process of providing additional personnel and assets for U.S. forces in Afghanistan, but the impact may not become apparent until later in 2009.

Language capability is also essential, in order to support the ability of U.S. forces to follow the key counterinsurgency injunction to live with the population, and requirements will grow proportionally with increases in U.S. ground forces in Afghanistan. The need is complicated by the fact that Afghanistan's two official languages – Dari and Pashto – are not mutually intelligible, and many Afghans know one but not the other. One option is to utilize U.S. servicemembers with local language ability, but such troops may be in short supply. Dari and Pashto are regarded as difficult languages to learn, requiring time to develop the ability to communicate in either of them on substantive matters. Another option is to utilize Afghan interpreters. One challenge is that, in addition to the general challenge of sufficient supply, Afghan languages vary from region to region; in some situations, a non-local Afghan interpreter might be understood but nevertheless regarded with some suspicion as an outsider. Variations in regional dialect English-language instruction now available to members of the Afghan National Security Forces might also ease some mil-to-mil communication barriers, but will not directly help U.S. forces communicate with local populations.

Legal Basis for Presence of International Forces

Two separate sets of arrangements are in place, for ISAF and for U.S. forces deployed under U.S. command, to provide a legal basis for the presence of those forces in Afghanistan.

Legal Basis for U.S. Forces

In 2002 and 2003, U.S. Embassy Kabul and the Afghan Ministry for Foreign Affairs exchanged diplomatic notes, which together constituted a formal agreement. The notes, which remain in force, confirmed that military and civilian personnel of the Department of Defense shall be accorded a status equivalent to that of Embassy administrative and technical staff under the 1961 Vienna Convention on Diplomatic Relations. The notes also addressed freedom of movement, licenses, the wearing of uniforms, the use of vehicles, exemption from taxation, and imports and exports. They confirmed U.S. criminal jurisdiction over U.S. personnel.[60]

Some of the basic provisions of that exchange of notes were reconfirmed by a joint declaration signed by President Karzai and President Bush, in May 2005, in which the two countries committed themselves to a strategic partnership with the goal of "strengthen[ing] U.S.-Afghan ties to help ensure Afghanistan's long-term security, democracy and prosperity."

The Declaration confirmed the bilateral intent to work together closely on a range of activities including, in the security sector: ANSF training, security sector reform, counterterrorism operations, counternarcotics programs, intelligence-sharing, border security, and strengthening ties with NATO. The Declaration included the specific, practical commitment that U.S. military forces operating in Afghanistan would continue to have access to Bagram Air Base "and facilities at other locations as may be mutually determined," and that U.S. and coalition forces would continue to enjoy freedom of action to conduct military operations "based on consultations and pre-agreed procedures."[61]

Legal Basis for ISAF Forces

United Nations Security Council Resolutions (UNSCR) provide the legal basis for the presence of ISAF forces in Afghanistan. A December 2001 UNSCR authorized, under Chapter VII of the United Nations Charter, the establishment of ISAF to "assist...in the maintenance of security in Kabul and its surrounding areas."[62] That mandate was based on a specific appeal for such a force included in the December 2001 Bonn Agreement.[63] In January 2002, the Interim Authority of Afghanistan signed a Military Technical Agreement with the newly formed ISAF.

In October 2003, the UN Security Council authorized an expansion of the ISAF mandate to include supporting GIRoA in maintaining security outside Kabul and its environs, and providing security to support the accomplishment of other objectives outlined in the Bonn Agreement.[64] The current UN mandate extends the authorization of ISAF for a period of 12 months beyond October 13, 2008.[65]

GIRoA Concerns

Over time, the Afghan leadership has expressed interest in making sure that ISAF- and U.S.-led forces coordinate their operations with the ANSF and with each other. For example, the 2006 Afghanistan Compact, the basic framework for international community engagement in Afghanistan in all sectors, states that all "OEF counter-terrorism operations will be conducted in close coordination with the Afghan Government and ISAF."[66]

In August 2008, President Karzai called for a review of the presence of all foreign forces in Afghanistan and the conclusion of formal status of forces agreements.[67] He issued the call during the heated U.S.-Iraqi negotiation process aimed at achieving a status of forces-like agreement, and just after U.S. airstrikes in Azizabad, Afghanistan, had apparently produced a number of civilian casualties. In January 2009, GIRoA reportedly sent a proposed draft agreement to NATO, which outlined terms and conditions for the presence of NATO forces in Afghanistan.[68]

Security Situation

The year 2008 witnessed an increase in security incidents that led some observers to argue that the insurgency was gaining ground – that the Taliban was "winning" – while others argued instead that insurgent tactics were evolving. The insurgency remained a loose and sometimes internally fractious network of Afghans, supported by some outside help including the availability of safe haven across the border in Pakistan.

Security Trends: Characterization

In general, the security climate in Afghanistan has tended to follow cyclical patterns, based on the seasons. The spring poppy harvest season draws some workers-for-hire away from the insurgency; insurgent leaders, who profit from the poppy crop, support this pattern. The forbidding winter cold makes movement and many activities harder, and usually finds some insurgents recuperating across the border in Pakistan. The warmer spring weather provides an opportunity for insurgents to attempt operations. Given the cyclical patterns, changes in security trends are best evaluated by year-to-year rather than month-to-month comparisons.

Recent years, by all accounts, have witnessed an upswing in security incidents. Many practitioners date the growing violence from mid-2006, when NATO assumed security responsibility first for southern, and then for eastern Afghanistan. Minister of Defense Wardak, for example, noted that in 2006 the insurgents "came on in a big way," and suggested that their intent had been to weaken political will in NATO capitals.[69]

ISAF officials note that from 2007 to 2008, there was a 33% increase in the overall number of kinetic events. ISAF defines "kinetic" events to include attacks against Afghan or international forces, whether by improvised explosive device (IED), indirect fire, or direct fire; but not, for example, kidnappings or intimidation. IED events, the single largest cause of casualties, increased by 27%. In addition, attacks on GIRoA officials and facilities increased by 119%. Afghan civilian deaths, in turn, increased by between 40 and 46%.[70]

Anecdotally, mid-2008 witnessed several insurgent attacks that prompted significant international attention. On July 13, 2008, in Konar province, nine U.S. soldiers were killed when their temporary outpost was attacked by 150 or more insurgents. On August 18, 2008, Forward Operating Base (FOB) Salerno in Khowst province was attacked by suicide bombers, killing Afghan civilians. Also on August 18, a large organized group of insurgents ambushed French forces in Kabul province, killing ten.[71]

Security Trends: Evaluation

These developments led some observers to conclude that the balance had tipped in favor of the insurgency. A study by the Paris-based International Council on Security and Development, released in December 2008, concluded that "...the Taliban has been experiencing a renaissance that has gained momentum since 2005. The West is in genuine danger of losing Afghanistan."[72] In a December 2008 Op-Ed, former head of UNAMA Lakhdar Brahimi wrote: "The [Afghan] government is losing ground every day to insurgents and other outlaws who now control at least a third of the country."[73] After embedding with Taliban fighters, journalist Nir Rosen concluded that the Taliban was winning.[74]

ISAF officials explain the increased number of security incidents somewhat differently. They point out that the growing presence of "friendly" forces, including international troops and Afghan forces – including 11 new Afghan National Army battalions and 10,000 more ISAF troops, from November 2007 to November 2008 – allowed the conduct of more operations, and thus more contact with the enemy. They add that deteriorating control of the border areas of Pakistan provided insurgents with additional safe haven opportunities.[75]

ISAF commanders also stress the importance of evaluating insurgent attacks qualitatively, as well as quantitatively. By late summer 2008, under pressure from international and ANSF operations, insurgents abandoned large-scale, relatively conventional-style attacks, because they were suffering heavy losses. Insurgent groups moved instead toward small-unit operations, applying more asymmetrical and more sophisticated tactics. These included attacks on key lines of communication (LOCs) including roads and bridges, more use of IEDs, assassination attempts against Afghan civilian and military officials, and attacks against government facilities such as district centers and schools. Insurgents, officials suggest, are "getting more effective" at these asymmetric activities. The specific impact of insurgent targeting of LOCs is hard to measure, ISAF officials note, since no systematic measure is made of highway traffic.[76]

At the same time, ISAF officials assess that the Taliban and other insurgents, while gaining some greater tactical facility, are not guided by a single, coherent strategy – the leadership does not appear to be formulating and directing an overall master plan.[77]

At the end of 2008, senior officials in Afghanistan tended to hedge their bets when describing the security climate. UN SRSG Kai Eide noted: "We haven't lost…but we haven't won…"[78] A senior U.S. military official described it this way: "We're winning the fight, but not the war".[79] Some COIN theorists argue, in turn, that such uncertainty is not neutral – that in a COIN fight, "not winning" is tantamount to losing.

Characterizing the Insurgency[80]

While many observers use the term "Taliban" as a short-hand for the insurgency in Afghanistan, senior western officials in Afghanistan stress that the insurgency is not unified. ISAF prefers the term "insurgent syndicate" to refer collectively to all its various strands. Further, insurgent activities are closely linked with criminality, always a potent force in ungoverned spaces, and in particular with drug cultivation and sales.

Taliban

The Taliban itself, Afghan and ISAF officials note, is more a network than a single organization.[81] The Taliban emerged from the Afghan civil war of the early and mid-1990's, and the organization ruled Afghanistan from its capture of Kabul in 1996 until its defeat in 2001. Mullah Mohammed Omar, the *de facto* head of state during Taliban rule, is generally assumed to be alive and leading the organization from Pakistan. In December 2008, for example, he reportedly issued new threats over the Internet against international forces in Afghanistan.[82] The Taliban leadership includes two main "shuras" (councils) – a leadership council in Quetta, Pakistan, under Mullah Omar's watch, and another shura based in Peshawar, Pakistan.[83] The Taliban reportedly receives support from some current and/or former Pakistani officials, including members of the Inter-Services Intelligence Directorate (ISI), in the form of logistics, medical, and training assistance.[84]

Haqqani Network

The Haqqani network is closely associated with the Taliban and one of its strongest factions. Reportedly, the network is also particularly closely linked to al Qaeda. Jalaluddin Haqqani fought as a mujahedin leader against Soviet forces, receiving substantial assistance from the CIA by way of Pakistan's ISI.[85] When the Taliban came to power, he joined the government as a Minister but retained a separate power base in his home Zadran district and tribe, east of Kabul. His son Sirajudin has reportedly ascended to a key leadership role, and has reportedly called for changes in the leadership of the Quetta shura. U.S. officials in

Afghanistan note that Sirajudin, like his father, has focused on his home Zadran district but has also expanded his activities into the areas south of Kabul.

Hezb-i-Islami Gulbuddin (HiG)

Gulbuddin Hekmatyar was a key mujahedin leader against Soviet forces. His organization, then known as the Hezb-e-Islami, received substantial aid from the U.S. Government, which reportedly considered him a key ally. He twice held the title of Prime Minister during the early 1990's civil war period, before seeking refuge in Iran when the Taliban came to power. He has reemerged in Afghanistan as the leader of the insurgent group, Hezb-i-Islami Gulbuddin (HiG), which is affiliated with both the Taliban and al Qaeda. In 2008, Hekmatyar apparently opened the door to talks with GIRoA, in part through a spring 2008 letter addressed to President Karzai. Some practitioners and observers suggest that there may be good potential for drawing Hekmatyar away from the insurgent fight and into a constructive role.[86] Others caution that his reputation for Islamic extremism and human rights abuses call into question the likelihood and advisability of any reconciliation with him.

Foreign Groups

Foreign groups play critical roles in the insurgency by variously supporting and enabling Afghan insurgents.[87]

Al Qaeda, which both enabled and leveraged the Taliban during its years in power, reportedly mobilizes foreign fighters from the Arab world, Chechnya, Uzbekistan, and other locations, to join the fight in Afghanistan.

Tehrik Taliban-i Pakistan (TTiP) is an umbrella organization for indigenous Pakistani Taliban commanders, based in Pakistan, in the Federally Administered Tribal Areas (FATA) along the border with Afghanistan. TTiP is led by Baitullah Mahsud, who is from South Waziristan in the FATA, and who has reportedly built up strongholds in North and South Waziristan by recruiting and training young men, and "killing uncooperative tribal leaders."[88]

Lashkar-e-Tayba, a Pakistani insurgent group originally focused on the disputed Kashmir region, reportedly cuts its insurgent teeth along the border with Afghanistan by training insurgents to fight there.

Tehrik Nefaz-e Shariat Mohammadi (TNSM) is a Pakistani insurgent group based primarily in the Northwest Frontier Province next to the FATA. Focused primarily on deepening its local control, the TNSM has also supported some Taliban operations in Afghanistan.[89]

Popular Support for the Insurgency?

The population is the center of gravity – the primary focus – in counterinsurgency operations. Evaluating a population's support, tacit as well as active, for an insurgency, as well as its perceptions of the fight, is one helpful tool for assessing the strength of that insurgency.

In general, it is common for insurgents to try to shape popular perceptions. Some experts argue that in 2008 in Afghanistan, the Taliban and other insurgents used high-profile attacks in and near Kabul to sow fear, or create a "sense of siege."[90] Those attacks included, among others, a strike against the opulent and popular Serena Hotel in January, an assassination attempt against President Karzai in April, a bombing at the Indian Embassy in July, and the kidnapping of International Rescue Committee workers in August.

Meanwhile, in late 2008, ISAF officials assessed that active Afghan popular support for the Taliban and other insurgents was not increasing – not least because the Taliban's ideology had little appeal for most Afghans. At the same time, ISAF assessed that popular support was shifting away from the Government of Afghanistan toward "fence-sitting", driven by frustrations with poor governance, lack of economic progress, corruption, and lack of security. COIN theory underscores the importance of at least tacit popular support for the government – that popular neutrality is insufficient to defeat an insurgency.

COIN Operations

At least as important to the success of the counterinsurgency as the number of Afghan and international security forces, experts argue, is what those forces actually do. As a rule, counterinsurgencies share an emphasis on "population security", but circumstances, and therefore the most effective approaches, may vary widely from one COIN campaign to another, or even geographically or temporally within a given COIN campaign.[91] In Afghanistan, COIN efforts have been challenged by especially rugged terrain, by limited forces and resources, and by the need to gain sufficient understanding of local areas to develop situation-specific approaches.

Shape, Clear, Hold, Build

In Afghanistan, COIN terminology, if not yet practice, has become a lingua franca shared by Afghan and international practitioners. The basic tenet of COIN operations in Afghanistan, borrowed and adapted from Vietnam and more recently Iraq, is "shape, clear, hold, build." In general, that approach includes working closely with Afghan counterparts to target insurgents, kinetically if so required; using Afghan or if necessary international security forces to hold an area once it is cleared; and applying coordinated civil-military efforts to begin building institutions and services once the security landscape in an area so permits.

COIN efforts in Konar province in eastern Afghanistan, along the border with Pakistan, are frequently presented as an instructive model of coordinated civil-military action. There, a U.S. Army battalion working with a very capable Afghan National Army (ANA) unit first cleared the area, and then established outposts to provide presence. U.S. forces, supported by a Provincial Reconstruction Team, then negotiated deals with local tribal shuras – if the shuras would provide security, they would receive economic development "in the form of roads, bridges, schools and health clinics." Central to the approach was the construction of a paved road – a rarity outside big cities in Afghanistan – in which local residents had a vested interest. The road gave forces the access they needed to secure village populations, it made it harder for insurgents to emplace IEDs, it gave civilian assistance agencies freedom of movement, and it gave local residents a critical tool to support economic development. The U.S. Commanders Emergency Response Program (CERP) funded payments to locals to provide security for the road.[92]

Practitioners stress, however, that there is no cookie-cutter model for COIN in Afghanistan – roads, for example, cannot always play the role they did in Konar province. In mid-2008, insurgents stepped up their attacks against Afghanistan's major road arteries, include Highway 1 that connects Kabul with Kandahar. Officials speculated that the insurgents attacking the highway in Zabul wanted to tie up the ANA and other forces, forcing them to build check points and leaving them fewer resources to focus on insurgent strongholds. In southern Zabul province, along the Highway 1 artery, military and civilian officials wondered if the "Konar model" approach to roads might be applicable. What they discovered, however, is that roads served a different function in Zabul – the highway was used primarily by inter-state truck drivers and the insurgents themselves, while local residents required only some passable means for getting to and from local markets. In other words, local residents did not perceive that they had the same, strong vested interest in the security of the main highway as residents of Konar province had in their own road.

Operations by the U.S. Marines in 2008, in the town of Garmsir, in southern Helmand Province, also reflected some key COIN approaches including an emphasis on population security, close cooperation with local security forces, the use of both kinetic and soft tools, and the incorporation of civilian resources. By the time operations commenced in April 2008, Garmsir had become a key insurgent transit and logistics hub, and much of the local population had withdrawn. The action began with a large-scale aerial insertion, followed by a month of clearing and fighting against small but tenacious and well dug-in insurgent groups. As the insurgents were defeated in a given area, the local population began to return, and the Marines increased their focus on population security, including the use of "gated communities" for population control, and reaching out to community leaders. As the fighting wound down, the ANSF joined the Marines on patrols to provide presence and "hold" the area. In fall 2008, the Marines handed security responsibility for the Garmsir area over to UK counterparts and the ANA, to continue the "hold" mission. From the planning stages onward, the Marines worked with the U.S. Agency for International Development, asking, "What do we want this to look like, afterwards?" The AID representative working with the 24th MEU was ready to initiate projects as soon as the security situation so allowed.[93]

ISAF commanders stress that through 2008, insufficient international and Afghan forces were available to shape, clear, hold, and build effectively throughout Afghanistan. The ISAF 2009 winter campaign, for example, conducted in selected districts chosen to include the majority of the Afghan population, aimed to "hold" ground and deepen security there, without significantly expanding GIRoA's control geographically.

LivingAmong the Population

As a rule, the center of any counterinsurgency is the population, and living among the population is a central tenet of COIN. As CENTCOM Commanding General, General David Petraeus, and many others have stressed, "You don't commute to work."[94] Senior officials at U.S.-led RC-East noted that as of November 2008, RC-East had between 130 and 140 combat outposts (COPs) in their area of responsibility (AOR) – "We're really out with the people." Most of those COPs were co-located with ANA counterparts or supported nearby Afghan observation posts, although a few were U.S.-only.[95]

For some Allies, constrained by national caveats, the premise of living among the population is a challenge. The idea is generally understood to mean getting outside the wire and interacting with the population as much as possible, not merely living adjacent to it. Allied forces that are largely confined to forward operating bases (FOBs) cannot follow the approach.

For ISAF in general, a major constraint on living with the population has been insufficient international forces. As one commander argued, "The strategy doesn't work when you don't have enough forces to do it."[96] Writing about U.S. efforts in both Afghanistan and Iraq, leading defense expert Colonel H.R. McMaster noted: "Decisions against deploying coalition forces in numbers sufficient to secure populations left many commanders with no other option than to adopt a raiding approach to counterinsurgency operations – an approach that tended to reinforce the perception of coalition forces as aggressors and conflated tactical successes with actual measures of strategic effectiveness."[97]

It might seem that by scaling back the size of the force presence at each outpost, the number of outposts could be increased and the overall footprint of forces "living with the population" could be extended. The problem is that, according to commanders on the ground, it takes one platoon to secure an outpost, so getting forces outside the wire to interact with the population requires two platoons at each location. As of November 2008, RC-East noted that about 40 of its COPs were single platoon-sized, with 30 Soldiers or fewer.

Afghanistan itself presents some particular challenges to living among the population – the Afghan population is much more rural and dispersed than that of Iraq, and many Afghans live in remote, isolated, and barely accessible valleys. Determining how best to live among the population may also sometimes require some discretion. Some practitioners and experts content that in many long-isolated valleys, the local population is generally hostile to all outsiders, whether "friendly" or otherwise. Insisting on living among them create new tensions, without necessarily strengthening that population's resistance to outside insurgents.

Borders

In general, successful counterinsurgency requires a closed system, so that COIN can gradually smother the insurgency. As discussed below, Afghanistan's open border with Pakistan significantly complicates the COIN effort. The border marks the boundary of the system for Afghan and international security forces, constrained by international law, but not for insurgents who make ample use of safe haven and resupply in Pakistan.

Role of Special Operations Forces in COIN

Special operations forces (SOF) play an essential role in COIN in Afghanistan, through direct action against insurgent leaders. One senior U.S. commander underscored SOF's "continuous disruptive effect on leadership." SOF efforts in Afghanistan include significant Allied as well as U.S. participation, as well as a growing role for elite Afghan "commando" forces.

The partnership between U.S. SOF and conventional forces in Afghanistan may differ somewhat from the analogous partnership in Iraq, due to the disparity in the size of the conventional force presence. In Iraq, especially in 2007 and 2008, a large conventional ground forces presence with a widely distributed footprint was able to gain substantial, detailed situational awareness about local conditions, and to provide that insight to SOF for use in targeting, and in planning operations. Further, the much larger conventional presence in Iraq made forces readily available to "hold" a given area once SOF cleared it. The significantly smaller conventional forces presence in Afghanistan may not yet have allowed the development such a robust SOF/conventional synergy.

Role of Air Power in COIN

The major combat phase of operations in Afghanistan relied heavily on ground SOF calling in airstrikes on al Qaeda and Taliban targets, and many experts pointed to those operations as a model of jointness – the ability of Military Services to work together seamlessly. Since then, due to Afghanistan's forbidding, mountainous geography, and to the relative dearth of Afghan and international ground forces, the COIN campaign has continued to rely greatly on the use of kinetic air power. A larger ground forces presence may reduce some of that reliance, but the terrain alone is likely to make air strikes a necessary counterpart to ground-based fires.[98]

Civilian casualties resulting from air strikes in Afghanistan have prompted strong expressions of concern from Afghan and international audiences. President Karzai strongly condemned a coalition air strike conducted on August 22, 2008, against the village of Azizabad, in Herat Province in western Afghanistan, which he said killed 95 Afghan civilians. Further complicating the incident were initial reports from the coalition that apparently conflicted with the numbers claimed by Afghan government officials and local residents.[99]

UNAMA further raised the stakes by issuing an early statement claiming "convincing evidence" that at least 90 civilians, most of them children, had been killed.[100] UN SRSG Kai Eide later asserted that he had been correct to strongly underscore the UN's grave concern with civilian casualties.[101] In a December 2008 interview, Eide warned again of the need to conduct military operations with care, and to guard against civilian casualties, stating, "I am not convinced that we are" listening to the concerns of President Karzai and the Afghan people.[102]

In September 2008, GEN McKiernan issued an ISAF Tactical Directive, which replaced a directive issued by his predecessor General Dan McNeill in June 2007. The new Directive stressed "proportionality, restraint, and utmost discrimination in the use of firepower." It provided specific conditions for the use of air-to-ground munitions, and underscored the need to minimize the risk to civilians.[103]

Regional Context

Afghan officials, and international practitioners and observers, generally agree that Afghanistan's security is intimately linked to its relationships with its neighbors, first of all Pakistan, and to relations among those neighbors. GEN McKiernan stated in late 2008, "This is a regional insurgency and it requires regional solutions."[104] GEN Petraeus added in early 2009, "In fact, those seeking to help Afghanistan and Pakistan need to widen the aperture even farther, to encompass at least the Central Asian states, India, Iran and even China and Russia."[105] By the end of 2008, most U.S. strategists had concluded that to be successful, a strategy for "Afghanistan" would need to address the broader region.

Pakistan[106]

A major challenge to the counterinsurgency effort in Afghanistan is the fact that the Afghanistan-Pakistan border is largely porous, and insurgents fighting in Afghanistan have long relied on safe haven and other forms of support in Pakistan. As a rule, counterinsurgency efforts assume a "closed system", in which persistent COIN efforts, and growing popular support, can gradually smother an insurgency, but Pakistan's open border disrupts that premise by giving Afghanistan's insurgents a ready escape hatch.

The insurgency problem is complicated by the fact that the Government of Pakistan (GoP) has traditionally enjoyed only limited control over the Federally Administered Tribal Areas (FATA) that border Afghanistan.[107] The FATA is a legacy of British rule. To boost the border defenses of British India, the British gave semiautonomous status to tribes in that area by creating tribal "agencies," largely responsible for their own security. The area became the "FATA" after independence. Regional experts Barnett Rubin and Ahmed Rashid have argued that today, the area is used as a "staging area" for militants preparing to fight in both Kashmir and in Afghanistan.[108]

Pakistan's turbulent recent history may further complicate the GoP's efforts to achieve control. That history has included the assassination of politician and former Prime Minister Benazir Bhutto in December 2007, just ahead of scheduled general elections. In February 2008, parliamentary elections brought to power a coalition of former opposition parties including Bhutto's Pakistan People's Party, led by her widower Asif Zardari. In August 2008, General Pervez Musharraf, who had come to power in a military coup in 1999, resigned as President of Pakistan. In September 2008, Zardari was elected president, completing a transition to civilian-led rule. The ability and will of that civilian-led government to exercise authority over Pakistan's security forces, and to take steps to stop insurgent activities, is not yet completely clear.

Throughout its short history, Pakistan has had deeply vested interests in Afghanistan. The international border – the British-drawn Durand Line – cuts through territory inhabited, on both sides, by ethnic Pashtuns, with significantly more Pashtuns living in Pakistan than in Afghanistan.[109] The Pashtun population of southern Afghanistan provided the primary base of support for the Taliban during it rise. Further, most observers underscore that the Government of Pakistan has a general interest in ensuring that Afghanistan is a regional ally, in part as a balance against Pakistan's long-simmering conflict with neighboring India. That broad interest was reflected in Pakistani support for the Afghan mujahedin fighting the Soviet occupiers in the 1980's, and later, for the Taliban regime – relationships that have created difficulties in post-Taliban bilateral relations.

In recent years, the GoP attempted to achieve a measure of stability along the border with Afghanistan by following the example of the British Raj and striking a series "truces" with local power brokers, in 2004, 2005, and 2006. In February 2005, for example, the Pakistani military reportedly reached a peace deal with Baitullah Mahsud, leader of the umbrella organization Tehrik Taliban-i Pakistan (TTiP), and withdrew its forces from check points in the region. In mid2006, Islamabad struck a major peace deal with insurgents in North Waziristan, agreeing to end military operations and remove local checkpoints, in return for an end to insurgent attacks on government officials.[110] In early- and mid-2008, Pakistani forces, tried a similar approach, pulling back from TTiP's stronghold in Waziristan in the FATA.[111] By all credible accounts, these "deals" did not lead to greater stability.

In July 2008, the U.S. Government reportedly confronted Pakistani authorities with evidence of ties between members of ISI and the Haqqani network in the FATA. At that time, President Bush authorized U.S. military cross-border operations into Pakistan, by ground or Predator unmanned aerial vehicles (UAV). Observers counted at least 11 attacks by Predator UAV in August and September 2008, in addition to a ground attack in early September.[112]

ISAF officials have noted that cross-border attacks have yielded big operational and tactical benefits – by causing the insurgent networks to feel disconnected, and prompting local residents in Pakistan to want al Qaeda and other outsiders to leave their communities.[113] At the same time, U.S. civilian and military officials acknowledge that such cross-border strikes have the potential to spark local protest and to destabilize the fragile Government of Pakistan, still struggling to consolidate civilian rule.

To be clear, NATO's policy for ISAF does not include cross-border strikes. Asked in July 2008 whether the Alliance would go after militants in Pakistan, Secretary General Jaap de Hoop Scheffer said, "My answer is an unqualified 'no'. We have a United Nations mandate for Afghanistan and that's it. If NATO forces are shot at from the other side of the border, there is always the right to self-defense but you will not see NATO forces crossing into Pakistani territory."[114]

By mid- to late-2008, efforts by the Pakistani military to tackle the insurgency problem had increased noticeably; senior U.S. officials and tactical-level military commanders in Afghanistan attributed the changes to the pressure from U.S. cross-border attacks. In August 2008, the Pakistani military stepped up operations in Bajaur, the northernmost of the seven agencies in the FATA, across from Afghanistan's Konar province. ISAF officials with access to imagery noted that after the operations, Bajaur resembled Fallujah, Iraq, after kinetic coalition operations in November 2004 – that is, with some allowances for the more rural setting in Pakistan, destruction from the relatively heavy-handed Bajaur operations was considerable. According to ISAF officials, while the Pakistani operations suggested some room for improvement in the "soft" skills of COIN, they had an impact by disrupting insurgent networks.[115]

In 2008, Pakistan's counterinsurgency approaches in the FATA also included arming "lashkars" – militias – of tens of thousands of local residents, to help maintain security.[116] The lashkars, which draw on the strength and local authority of traditional tribal structures, were intended to help "hold" areas after they were cleared by military operations.

Late 2008 also witnessed an improvement in cross-border coordination. GEN McKiernan noted that the Tripartite Commission, a senior-level U.S.-Afghan-Pakistani coordination body, had resumed its functions and convened three times between June and November 2008.[117] RC-East reported that cooperation among Pakistani, Afghan, and U.S. forces continued to grow at the trilateral Border Coordination Center (BCC) at the Torkham Gate, one of six planned BCCs. That collaboration benefited from Predator feeds that provided a common picture of the battlespace. ISAF and their Afghan counterparts planned to establish 6 BCCs by Fiscal Year 2010.[118]

At the tactical level, U.S. ground forces in eastern Afghanistan reported that in late 2008, the tenor of their regular tactical-level border coordination sessions grew more constructive. Tactical-level coordination improved – including, by November 2008, 13 cases of direct cross-border coordination with Pakistani forces, to "fix and defeat the enemy at the border," particularly along the border with Afghanistan's Paktika province.[119]

Overall, senior U.S. officials in Afghanistan, and outside observers, suggest that substantial improvement of the Afghanistan-Pakistan border situation will require from Pakistan both political will and appropriate capabilities. GEN McKiernan stated in November 2008 that he has seen "a shift in thinking at the senior levels in Pakistan that this insurgency is a problem that threatens the very existence of Pakistan."[120] Other U.S. commanders in Afghanistan noted, in late 2008, that they have observed a "sea change" in the views of

Pakistani military officials, who increasingly view insurgents as existential threats, and who are attempting to improve the COIN capabilities of Pakistani forces to counter the insurgents. Remaining differences, U.S. military officials suggest, tend to include different perceptions of various insurgent groups and the threats they represent.[121] Other observers suggest that the extent of the commitment of senior Pakistani civilian officials to defeat the insurgent challenge is less clear.

Observers suggest that a key strategic question concerns the fragility of the Pakistani polity, and the extent to which the GoP can be encouraged or pushed to cooperate in counterinsurgency efforts without significant domestic political blowback. For example, in October 2008, the Pakistani parliament unanimously passed a resolution calling for an end to military action against extremist groups, and its replacement with dialogue. The resolution stressed the need for an "independent foreign policy" for Pakistan, and stated that "the nation stands united against any incursions and invasions of the homeland."[122]

Iran

ISAF officials note that the role of Afghanistan's large neighbor to the west, Iran, is also critical to its future, and they describe Iran's approach as a "dual-track strategy." On one hand, Iran enjoys close, long-standing cultural, linguistic, and religious ties with significant portions of Afghanistan's population. ISAF officials estimate that Iran is the second-largest contributor of reconstruction assistance to Afghanistan, after the United States – its efforts are most evident in Herat Province in western Afghanistan. And since Iran is a major destination for Afghan heroin, with all of its attendant concerns about crime and drug addition, Iranian officials share with their Afghan counterparts a vested interest in effective counternarcotics approaches. Some officials also point to the generally positive role Iran played at the 2001 Bonn Conference, to help forge consensus among Afghan factions about the creation of a post-Taliban government, as evidence that Iran can play a constructive role on Afghan matters.[123]

At the same time, ISAF officials state that Iran has provided some weapons and training to Afghan insurgents. Some add that Tehran may be concerned about a growing U.S. military footprint along both its eastern and western borders, as additional U.S. military forces flow into southern Afghanistan, and U.S. forces assume battlespaces in southern Iraq that were formerly manned by coalition partners. One official argued that Iran's interest is to "keep it simmering" in Afghanistan.[124] Most practitioners and observers suggest that, in some capacity, a comprehensive solution for Afghanistan must take Iran into account.

Afghan National Security Forces (ANSF)[125]

A fundamental premise of counterinsurgency operations is the central importance of host-nation forces – including establishing and improving those forces should their quantity or quality be insufficient. Decades of war, displacement, and mismanagement, followed by the defeat of the Taliban regime, left Afghanistan without organized, functioning security forces or equipment, so rebuilding the Afghan National Security Forces (ANSF) has been a high priority of the post-war international assistance effort. ISAF's broad goal is to transition "lead security responsibility" to the ANSF who are focused, as a rule, on the current fight, not Afghanistan's long-term security requirements. From a security perspective, one positive legacy of the years of conflict may be the fighting spirit so common among many Afghans, acknowledged by the common refrain of international military officials: "They will fight!"

Afghan National Army (ANA)

All of the Afghan security forces are still developing, but the Afghan National Army (ANA), under the Ministry of Defense, is currently, by a wide margin, the most capable force.

ANA Numbers

As of December 2008, approximately 68,000 soldiers were assigned to the ANA, of around 150,000 total assigned ANSF.[126] That marks an increase from 29,366 in July 2006, 39,081 in July 2007, and 65,547 in July 2008.[127]

The Bonn Agreement established an endstrength target of 70,000 for the ANA. A decision was made in early 2008 to stretch that goal to 86,000. In September 2008, the Joint Coordination Monitoring Board (JCMB) – the body co-led by GIRoA and UNAMA, and charged to oversee implementation of the Afghanistan Compact – endorsed GIRoA's plan to increase that target figure to 134,000. The new total would include a base of 122,000, plus a personnel "float" of 12,000, on the model of the U.S. and other militaries, to support ongoing training and education.

U.S. military officials training the ANA note that the target date for reaching the 134,000 total is the end of 2011, based on a growth rate, as of late 2008, of 28,800 per year, with an estimated attrition rate of 8 to 9%. In theory, one option for accelerating the training of new Afghan soldiers might be boosting the monthly intake of new recruits. One challenge would be the constraints of the training process, including the capacity of the Kabul Military Training Center (KMTC), and the numbers and capabilities of Afghan trainers.[128]

Senior U.S. practitioners in Afghanistan almost universally question whether the ANA target of 134,000 is sufficient to meet the security challenges Afghanistan faces. In November 2008, GEN McKiernan noted that 134,000 is "probably not the right number," and the outgoing head of the Combined Security Transition Command-Afghanistan, Major General Robert Cone, said that "134,000 probably isn't enough." Another U.S. commander in Afghanistan called the 134,000 target a "blip in the road", and still another estimated that altogether, the ANSF would need to grow to 300,000.[129]

For their part, many Afghan officials share the view that "134,000" will prove insufficient. Minister of Defense Wardak stated that Afghanistan had never yet had the proper proportion of troops to the area to be secured and to the population to be protected. Current force sizing, he noted, assumes the presence of a large international force – which will not always be there, and whose capabilities, he argued, are roughly double that of their Afghan counterparts. He concluded that "between 200,000 and 250,000 would be the proper size for the ANA."[130] In a September 2008 analytical report, a former civilian advisor to President Karzai argued that the ANA endstrength should be "at least 220,000."[131]

Some outside experts have also strongly advocated programming for a larger ANA. COIN theorist John Nagl, who has helped train U.S. personnel to train the ANSF, argued in November 2008 that the ANA should grow to 250,000.[132]

Force Modernization

In 2008, the focus of the ANA itself, and of the coalition ANSF training and advisory effort, was counterinsurgency, with an emphasis on the equipment that could be most readily fielded, and the skills that could most readily be developed and applied to the fight.

One shift, in late 2008, was initial "NATO-ization" of some ANA weapons, first of all a shift from the AK-47 assault rifle of Soviet origins, to the M-16 rifle of U.S. origin, widely used by many NATO countries. The AK-47 was a natural choice, as a starting point – years of Soviet sponsorship and then occupation had made the weapons widely available throughout Afghanistan. In November 2008, the first shipment of M-16's arrived in Kabul, and U.S. trainers began training Afghan army trainers. Some critics have noted that the M-16 is more temperamental to use and maintain than the AK-47, and that it will require an adjustment for Afghan forces. Supporters underscore the operational importance of Afghan interoperability with NATO counterparts.

As of late 2008, warfighting ANA Corps commanders were focused exclusively on the current fight – making sure that their soldiers had the equipment and training necessary for counterinsurgency.[133] Minister of Defense Wardak, however, had taken a longer-term look at possible future requirements, including the traditional military role of providing external defense. That outward-looking perspective was reflected in his suggestion to size the ANA by comparing it with the armies of Afghanistan's neighbors – Pakistan, Iran, and "the bear to the north". To balance between current and future requirements, he urged equipping the ANA "with a mix, right from the beginning, so it works for COIN and later on". Afghanistan needs a force that is "light but as effective as heavy forces", he added, and should include tanks, and an infantry combat vehicle – protected mobility with some firepower.[134]

ANA Structure and Organization

In contrast to the post-war Iraqi army, which was built from the ground up starting with small units, the ANA has been built from the top down, starting with headquarters leadership and staff, and then gradually fielding units under those headquarters.

As of the end of 2008, the ANA had five ground forces Corps Headquarters – the 201st Corps in Kabul, the 203rd Corps in Gardez in the east, the 205th Corps in Kandahar in the south, the 207th Corps in Herat in the west, and the 209th Corps in Mazar-e Sharif in the north. Under those Corps headquarters, the ANA had 19 brigade headquarters, and 71 "kandak" (battalion) headquarters.[135] As of January 2009, according to ISAF officials, the ANA had 46 kandaks capable of battalion-level operations.[136]

ANA "Corps" follow the European model, in which a Corps is a two-star headquarters, whose subordinate units are brigades – much like a U.S. Army Division. The five ANA Corps areas of responsibility (AOR), like the ISAF Regional Commands, are situated in the center, east, south, west, and north, but the ISAF and ANA boundaries do not completely correspond.

ANA Operations and Capabilities

By the end of 2008, all ANA Corps were engaging regularly in combined operations with ISAF counterparts. The overall percentage of deliberate combined operations that were ANA-led had increased from 49%, in the period from October to December 2007, to 62% in the period from July to September 2008. The extent of ANA leadership of such operations varied significantly, however, among ISAF Regional Commands, from ANA leadership of 23% of combined deliberate operations in RC-South, to 43% in RC-North, to 79% in RC-East, and 86% in RC-West. U.S. officials responsible for training estimated that the disparities might have more to do with variations in the coalition-Afghan partnerships, from region to region, than with variations in the capabilities of the ANA Corps or security conditions in the various AORs.[137]

In qualitative terms, ISAF officials note that ANA operational capabilities grew markedly during 2008. For example, in June 2008, following a prison break in Kandahar, initiated by a suicide bomber, the ANA deployed more than 1,000 soldiers south from Kabul, providing over half of the air lift required to transport them, within 24 hours. RC-East officials noted in November 2008 that in the previous month, they had participated in 35 combined "air assault" missions with the ANA, most of them ANA-led. In one mid-November action in eastern Afghanistan, for example, a combined ANA, ISAF and Afghan police force air assaulted into an area to be cleared. There, the Afghan police knocked on the doors, the ANA provided the inner cordon, and ISAF forces provided an outer cordon.[138]

Meanwhile, coalition special operations forces are helping the ANA develop elite "commando" kandaks. In theory, the commando forces would be capable of working with coalition counterparts on high-value targeting lists, and also of playing key roles in broader ANSF COIN operations.[139]

ANA as a National Institution

A number of observers have suggested that the ANA may be Afghanistan's only truly "national" institution. The outgoing head of CSTC-A noted in late 2008 that the ANA was "very integrated". In the immediate post-Taliban years, ethnic Tajiks and Uzbeks – strongly represented in the Northern Alliance – predominated in Afghanistan's fledgling army, at the expense of ethnic Pashtuns. But as of the end of 2008, the ANA ethnic balance more closely corresponded to that of the population of Afghanistan – Tajiks, about 27% of the population, accounted for between 30 and 40% of the ANA, while Pashtuns, 42% of the population, made up 41% of the ANA. One caveat is that ethnic balance may not always correspond to geographic balance – for example, instead of recruiting Pashtuns from former Taliban stronghold areas in southern Afghanistan, the ANA may look to Pashtun communities in other parts of the country to achieve balance.[140]

Meanwhile, Minister of Defense Wardak underscores that the ANA is well-regarded by the Afghan population.[141] This claim was supported by the results of a major survey of popular opinion conducted in 2008, under the auspices of the Asia Foundation, which identified the ANA as the public institution enjoying the highest level of public confidence in Afghanistan.[142]

Key Challenges to ANA Development

While ANA operational capabilities, by all accounts, continue to grow, the Army continues to contend with critical gaps and challenges. Like the Iraqi Army, the ANA lacks sufficient enablers, including logistics; intelligence, surveillance and reconnaissance (ISR); and air capabilities such as close air support (CAS). It continues to rely on U.S. and coalition forces for such support. In addition, the ANA faces a significant demographic gap, of personnel between the ages of 35 and 55, the legacy of Afghanistan's recent history of warfare. While the ANA can draw on its "older" personnel now to serve in leadership capacities, it will effectively take a generation to fully train and prepare the next contingent of

ANA senior leaders. Further, the ANA – like Afghan society as a whole – suffers from a ravaged supply of Afghan human capital. Since a significant majority of new recruits are illiterate, ANA training relies on methodologies that do not utilize written language; and a number of literacy instruction opportunities are available.

Afghan National Army Air Corps (ANAAC)

The Afghan National Army Air Corps (ANAAC), effectively Afghanistan's air force, is organizationally part of the ANA and is considered its 6th Corps. Afghanistan has an independent air force tradition dating back to 1924 – by the 1980's, after several periods of substantial Soviet assistance, Afghanistan had built a rather formidable air force. During the Taliban era, Pakistan assumed the foreign patronage role. During the war in 2001 that ousted the Taliban, Afghanistan's fleet was largely destroyed. Years of flying experience left the Afghans some human capital to draw on, in building a post-Taliban air force – although the current average age of its pilots, 44.7 years, is approximately the average life expectancy for Afghan males.

The ANAAC is trained and mentored by the Combined Air Power Transition Force (CAPTF), part of the CSTC-A. The CAPTF describes its ambitious goals for the ANAAC this way: "The ANAAC will be focused on the unique demands of Afghanistan but will also be modern, interoperable and sustainable, and integrated with the ANSF, capable of joint and combined operations…."[143]

CAPTF officials note that ANAAC development is proceeding in stages, based on agreements with ANSF leadership, with an initial emphasis on contributing to the COIN fight, first of all through air mobility. Afghanistan's unforgiving terrain and dearth of sufficient highway and rail transportation make the ability to move troops and supplies absolutely critical. Later – in the period between Fiscal Year 2011 and Fiscal Year 2015 – it is expected that the ANAAC will begin to acquire limited attack and ISR capabilities. Sometime thereafter, CAPTF officials note, the ANAAC might begin to build external defense capabilities, including air interdiction, but that is not a current focus. That timeline reportedly sits uneasily with some "legacy" Afghan fighter pilots, eager to rebuild the air force they once knew.[144]

As of November 2008, the ANAAC included 301 pilots, and the Afghan fleet included 31 tails: 18 Mi-17 and 3 Mi-35 helicopters; 6 AN-32 and 2 AN-26 transport fixed wing aircraft; and 2 L39 "Albatros" jet trainer aircraft. The ANAAC expected to receive additional aircraft by donation: 6 more Mi-35's, 3 in November 2008 and 3 in January 2009, from the Czech Republic; and 3 more Mi-17's from the United Arab Emirates in April 2009. The current fleet, and the donations expected in the near term, are all of Soviet-bloc origin – CAPTF officials note that the first priority was to acquire early capability by capitalizing on aircraft familiar to the Afghans. Plans call for shifting the fleet's orientation away from former Soviet technology in future acquisitions, including fixed-wing cargo aircraft.[145]

By the end of 2008, the ANAAC was making substantial contributions to Afghan and coalition COIN efforts. In October 2008, the ANAAC set new records by transporting 9,000 passengers and 51 tons of cargo, and by flying 908 sorties. At the beginning of 2008, according to CAPTF officials, ISAF met 90% of ANA transport requirements for cargo and passengers, but by November, the ANAAC was meeting 90% of the requirement.[146]

Afghan National Police

The Afghan National Police (ANP) are Afghanistan's civilian security forces, which fall under the Ministry of the Interior. The ANP includes several distinct forces: the Afghan Uniform Police (AUP), responsible for general policing; the Afghan National Civil Order Police (ANCOP), a specialized police force that provides quick reaction forces; the Afghan Border Police (ABP), which provides law enforcement at Afghanistan's borders and entry points; and the Counternarcotics Police of Afghanistan (CNPA), which provides law enforcement support for reducing narcotics production and distribution.[147]

According to ANP officials, the ANP are being developed as a paramilitary force to contribute to the counterinsurgency effort by joining the ANA in COIN operations, and by protecting the population after the ANA "clears." As of late December 2008, the ANP included approximately 68,000 assigned personnel.[148] The Bonn Agreement established a target ANP endstrength of 62,000; the current target endstrength is 82,000.

Police Corruption

The most commonly expressed concern of Afghan and international senior officials about the ANP is that they are not merely incompetent but also corrupt – they have, GEN McKiernan has stated, "a lack of basic policing values."[149] Some observers charge that such corruption is more than an obstacle to a job well done, in that it also alienates the population – the center of gravity in COIN – who may grow to see the Taliban as no worse than equally abusive civilian authorities.[150] Curiously, a major recent survey of Afghan popular opinion indicated that the ANP is the second most highly regarded public institution, after the ANA.[151]

Focused District Development

To address the problem, GIRoA and coalition forces launched the Focused District Development (FDD) initiative to retrain and reform local AUP forces, district-by-district. In the FDD program, the AUP are pulled out of a given district and sent to an intensive training course. Highly skilled ANCOP forces fill in, during their absence. After the AUP return, in order to reinforce their new skills, they operate under the tactical overwatch of, and then with mentoring by, coalition forces. As of January 2009, the AUP in 52 districts were undergoing the FDD process.[152] Coalition officials assess that FDD is generally successful, in that fewer violations by the AUP are reported after the training.

Some observers, including senior officials from international organizations, have charged that the program is not comprehensive enough to be effective. "Taking thugs away for a few weeks," one official observed, "just gives you better-trained thugs."[153]

Some outside observers, in turn, noting the urgent need for more and better policing on the streets of Afghanistan, have pushed for accelerated recruitment and fielding of weapons and equipment to the Afghan police. Coalition officials caution, however, that the reform process will take time, since the aim is a fundamental cultural shift. Providing gear, they argue, especially weapons, to "unreformed" districts, without proper accountability, would likely prove counterproductive.[154]

Afghan Border Police

By many accounts, the Afghan Border Police (ABP) may be beset by even greater incompetence and corruption than their AUP counterparts. To counteract these trends, GIRoA, working with coalition counterparts, launched the Focused Border Development (FBD) program, similar to the AUP's FDD. Between October 2008 and September 2009, FBD is scheduled to train 52 company-sized units of ABP, at four training sites. The courses are conducted by U.S. private security contractors – Blackwater and DynCorp. The retraining also includes arming the ABP with heavier weapons, including Soviet-origin DShK heavy machine guns.[155]

The FBD initiative, like FDD, relies on follow-up mentoring by coalition forces, after the completion of the formal training sessions. Those mentorship responsibilities are assigned to ISAF battlespace owners. As the outgoing head of CSTC-A observed, FBD is possible "because the 101st is helping me and giving me assets."[156]

Coalition and Afghan officials readily acknowledge the great challenge of securing Afghanistan's borders. Afghanistan has nearly 3,500 miles of borders, primarily in difficult, remote, mountainous terrain. Minister of Defense Wardak flatly observed, "We will never be able to secure the whole border."[157] Protecting the borders, some officials suggest, may require not only trained and professional ABP personnel stationed along the border, but also additional aerial reconnaissance and quick response forces.

Command and Control

Command and control arrangements for the ANSF have been adapted to current COIN efforts, which require "joint" action by multiple Afghan forces together with coalition counterparts. The Ministry of Defense and the Ministry of the Interior maintain formal command authority over their own forces – the ANA and the ANP, respectively.

To facilitate coordination, GIRoA created, in 2008, a series of Operations Coordination Commands, at the regional (OCC-R) and provincial (OCC-P) levels. There are 6 OCC-R's, one in each of the five ISAF Regional Commands, and one for Kabul city; and 34 OCC-P's are being established. OCC's at both levels are physical (not virtual) facilities that facilitate monitoring and coordination of operational and tactical-level operations.

OCC's include representatives from the ANA; the ANP; and the National Directorate of Security (NDS), Afghanistan's intelligence service. ISAF and CSTC-A provide mentoring. The command relationships among the participating organizations are purely "coordination", not "command". For example, as contingencies arise, OCC members provide direct conduits of information with their respective organizations – OCC-P members reach out to ANA brigades and ANP provincial command centers; while OCC-R members reach out to ANA Corps and ANP regional command centers. OCC-P's do not report to OCC-R's, and there is no national-level analogue. The ANA serves as the "lead agency" for OCC's, although OCC's may be physically located in police facilities.

Looking to the future, some observers have wondered how appropriate the OCC construct will prove to be for a "post-COIN" context when, for example, the focus of the ANA shifts from domestic to external concerns. A future transition might not prove especially difficult, since the OCC coordination relationships complement but do not replace the formal service command relationships.[158]

Training the ANSF

Since its inception, the international ANSF training effort has been characterized by multiple initiatives adopting sometimes divergent approaches, with a general trend toward greater unity of effort, and a stronger U.S. leadership role, over time.[159] Secretary of State Clinton has underscored President Obama's statement "that we must focus more attention and resources on training the Afghan Security Forces."[160]

Background and Organization of the Training Effort

The December 2001 Bonn Conference recognized the need for the international community to help the fledgling Afghan authorities with "the establishment and training of new Afghan security and armed forces." In early 2002, broad agreement was reach on a model in which individual "lead nations" would assume primary responsibility to coordinate international assistance in five different areas of security – these included placing ANA development under U.S. leadership, and police sector development under German leadership. The 2006 Afghanistan Compact transferred formal "lead" responsibility to GIRoA.

In 2002, to execute its "lead nation" role, the United States created the Office of Military Cooperation-Afghanistan (OMC-A) to train the ANA. In 2002, to supplement German efforts, the U.S. Government launched a police training initiative, led by the Department of State's Bureau of International Narcotics and Law Enforcement Affairs (INL), through U.S. Embassy Kabul, with contractor support. In 2005, the U.S. Government restructured its ANSF training efforts, shifting responsibility for supporting Afghan police development to the Department of Defense, and renaming the OMC-A the Office of Security Cooperation-Afghanistan (OSC-A).[161] Early in 2007, when the U.S. three-star military headquarters, the Combined Forces Command-Afghanistan (CFC-A) was deactivated, OSC-A was re-designated the Combined Security Transition Command-Afghanistan (CSTC-A), and assigned directly to US CENTCOM; CSTC-A was assigned to USFOR-A when that headquarters was established in 2008.

Training Teams

CSTC-A's primary mechanism for training and advising the ANSF is the use small teams that typically live and work with ANSF units. U.S. advisory teams working with the ANA – Embedded Training Teams (ETTs) – include between 12 and 20 personnel. ETTs work for CSTCA but are under the operational command of U.S. battlespace owners during combined operations with the ANSF. Non-U.S. NATO advisory teams are known as Operational Mentor and Liaison Teams (OMLT). In theory, their functions are similar to those of the ETTs, but due to national caveats, there are great variations in the degree to which OMLTs participate in operations with their Afghan counterparts. Advisory teams working with the ANP are known as Police Mentoring Teams (PMTs).

Some officials in battlespace-owning units have argued that the quality of ETT and PMT personnel varies and that, as one official put it, "the ETTs are better suited to planning than execution."[162] In October 2006, the U.S. Army consolidated, at Fort Riley, Kansas, the pre-deployment training and preparation of U.S. Army, Navy and Air Force personnel for assignments as advisors to Afghan and Iraqi security forces; in October 2008, the Army announced that the program would shift to the Joint Readiness Training Center at Fort Polk, Louisiana.[163] One reason for some reported variation in the quality of U.S. advisors may be that they are selected on an individual basis and come from a wide variety of backgrounds that may or may not include extensive operational experience.

Since its inception, officials argue, the training and advisory effort has been undermanned. According to CSTC-A, as of November 2008, advisory teams were manned, overall, at 64%. That marked an increase from 54% earlier in the year, before the 2nd Battalion, 7th Marines arrived and assumed police training responsibilities in addition to battlespace ownership. ANA teams were more fully staffed, at 78%, than ANP teams, at 33%. Those November 2008 figures reflected a re-balancing from earlier in the year, when ANA teams were resourced at 93% but ANP teams were only resourced at 12%. They also reflected some breaking down of available advisory teams into smaller pieces, in order to extend coverage to more Afghan units.[164]

As of late 2008, CSTC-A had requested 2,300 additional police advisors, and 60 additional ANA advisory teams.[165] Those ANA teams could come either from the U.S. military or from NATO – although NATO, despite repeated requests, has been reluctant to provide ANA advisory teams in sufficient numbers, or to provide embedded police trainers. To be clear, the requirement is not static – as ANSF units improve their capabilities, their partner advisory teams can be "harvested" and assigned to new ANSF units.

Mentoring the ANSF Leadership

In addition to tactical- and operational-level training, the coalition ANSF advisory effort also includes mentoring Afghan senior leaders – Ministers, senior ministry officials, and senior ANA and ANP commanders. ISAF and CSTC-A senior leaders invest considerable time in working closely with the senior leadership of the Defense and Interior Ministers, and of their regionally-based commands. President Karzai replaced the Minister of Interior in October 2008, a step favored by a number of senior U.S. and other international officials in Afghanistan, and widely viewed as an effort to curb corruption.[166] The coalition "advisory" role is strong – Afghan regional commanders regularly seek coalition support, and advocacy with their respective ministries, for identified requirements. The fundamental challenge is building institutional capacity – including leadership ability, physical infrastructure, effective systems, and trained and competent human resources.

Unit Partnering

In 2008 in Iraq, the Iraqi security forces (ISF) training and advisory effort relied on two complementary approaches, embedded teams and "unit partnering". Unit partnering involved matching a full coalition unit with an Iraqi counterpart of equal or greater seniority in a mentoring relationship. Multi-National Corps-Iraq directed the use of such partnerships where appropriate. Those relationships, in the words of U.S. commanders in Iraq, provided the opportunity to "show, rather than tell" – to provide visible examples of a competent unit and staff in action.

In Afghanistan in late 2008, the use of unit partnering was more ad hoc, and more contentious. Overall, unit partnering was less widespread than in Iraq, a natural consequence of a far lighter coalition footprint, since unit partnering requires available, locally-based units of appropriate size. Nevertheless, battlespace-owning U.S. units underscored the importance of such partnerships – as one commander noted, "ANSF capacity-building is our main effort, and we accept some risk in our operational capabilities to focus on this." For example, one U.S. brigade-sized Task Force sends its tactical command post including key brigade staff, for two weeks every month, to co-locate and partner with the nearest ANA Corps headquarters. Its Military Police (MP) battalion headquarters staff, in turn, work closely with the ANP Regional Command Center.

Further, the ANP re-training effort, for both AUP and ABP, requires substantial follow-up in the form of mentoring. Battlespace-owning units were tasked to establish and maintain those mentoring partnerships – a form of unit partnering – with the "reformed" ANP.[167]

At the same time, in late 2008, senior Afghan and CSTC-A officials evinced an antipathy toward the concept of unit partnering. Minister of Defense Wardak argued forcefully: "There is some talk that we should do partnering, but I am against it – our units are standing on their own feet. I will try very hard to push against this partnering. If they have partner units, they would lose their ability to learn and operate independently."[168] CSTC-A officials argued, similarly, that ANSF units tended to perform less well when partnered with coalition units, and that advisory teams were more effective than "partners" in encouraging the ANSF to take initiative.[169]

European Union Police Mission in Afghanistan (EUPOL)

In June 2007, the European Union (EU), recognizing a still-urgent need for competent Afghan police, established a police training mission in Afghanistan (EUPOL). As of December 5, 2008, EUPOL's strength stood at 177 international and 91 local staff, from 18 EU member countries plus Canada, Croatia, and Norway.

In general, EUPOL has a longer-term, more institutional focus than CSTC-A. Its mission statement declares: "The mission aims at contributing to the establishment of sustainable and effective civilian policing arrangements under Afghan ownership and in accordance with international standards."[170]

Some senior ISAF and Afghan officials criticize EUPOL for taking too long-term a perspective, for failing to work at the local level, and for focusing on steady-state, peacetime policing skills rather than on the paramilitary capabilities that, they argue, are urgently needed for the counterinsurgency effort.[171] UN SRSG Kai Eide, in turn, has noted that EUPOL staffing should increase significantly and has urged the EU to provide more – and more appropriate – personnel.[172]

Community and Tribal Outreach

In early 2009, Afghan and ISAF officials were finalizing plans for a new community outreach program, sometimes called the "community guard program", designed to take what GEN McKiernan has called a "bottom-up, community-based approach to security."[173] The premise is that neither international forces nor the ANSF have sufficient numbers to provide full population security, and are not likely to have them in the near future, even with expected increases in ISAF and ANSF troop strength. Minister of Defense Wardak stated in late 2008, "There is still a big gap between forces available, and the space to secure, so we need help."[174]

In Afghanistan, local community leaders are often tribal leaders, and local community structure is intimately linked with tribal structure, though not necessarily clearly or consistently. Observers describe tribal bonds as "pragmatic, localized allegiances," which may have been shaped over time by migration, competition for resources, reallocation of land rights as rewards for services, and links to the narcotics trade. While residents of several isolated valleys may all belong to the same tribe, their fiercest rivalries may be with fellow tribesmen in an adjacent valley. Fostering local community support for security initiatives generally involves working with tribal leaders, among others, but given Afghanistan's complex tribal affiliations, the risk of "getting it wrong" is relatively high.[175]

The community guard program attempts to avoid "getting it wrong" by focusing on the concept of "community outreach," rather than "tribal outreach". The initiative began with the recognition of the need to protect Highway 1, the key artery running south from Kabul to Kandahar and the site of escalating insurgent attacks in mid-to-late 2008. The program was expected to begin with a pilot project in Wardak and Logar provinces, just south of Kabul. Muhammad Halim Fidai, Governor of Wardak province, was quoted as saying: "We don't have enough police to keep the Taliban out of these villages and we don't have time to train more police – we have to fill the gap now."[176]

In the program, each local "community" – including all relevant tribes – would select representatives to a shura; the shura, in turn, would select project participants to help provide security, for example through neighborhood watch efforts and guarding fixed sites. One goal, U.S. officials noted, is that local residents adopt a "not in my village" attitude toward insurgents and criminals.[177]

Funding for the program would be provided by U.S. CERP funds; this funding would not cover arming participants. U.S. accountability requirements mandate formal U.S. oversight, but additional Afghan oversight would be provided by both on-site ANSF representatives, and the moral authority of the community shura. The goal – as U.S. Ambassador William Wood stated – is "to empower locals, both local governance and tribal structures, to make them work for themselves."[178] "The idea," as Minister of Defense Wardak expressed it, "is to bridge the gap between the government and the people, and to make the people feel responsible."[179]

The initiative draws to some extent on the model of "arbakai" – a traditional Pashtun institution, in which a tribally-based auxiliary force is formed to defend a village and its surrounding area on a temporary basis. That familiar association may help smooth the introduction of community outreach, but it also raises concerns in some quarters that the program might reignite and facilitate local warlordism. One U.S. senior official stated in late 2008, "I worry about arbakai."[180]

Minister of Defense Wardak is reportedly particularly concerned about the danger of a return to the chaos of the early 1990's civil war. He stated in late 2008 that the new program must not rearm anyone – "we should not create new warlords or reinforce old ones."[181] U.S. officials quoted a senior ANA commander as saying, reflecting cautiously on the initiative, "The army and police serve a nation, but a militia serves a man."[182]

The program must also overcome the legacy of similar and more recent initiatives to generate security at the local level. The Afghan National Auxiliary Police program was created in 2006, amidst some controversy, as a stop-gap measure in southern Afghanistan. The locally-recruited force, including many who previously worked for warlords, had an approved size of 11,271. Recruits were given ten days of training, and members received the same salaries as regular ANP street cops – $70 per month.[183] A number of practitioners and observers argued that the training was insufficient to produce a credible security force. At the time, the head of CSTC-A called the program "an attempt to take short-cuts," and its participants "a bunch of thugs," and more recently, an RC-East senior official concluded that they "went brigand."[184] By late 2008, the program had been completely dismantled.

Reconciliation

Military theorists and practitioners contend that war is a contest over the terms of an ultimate political settlement. In Afghanistan, both GIRoA and the various insurgent groups are contending to set the conditions for a final settlement.

In an early preview of a possible end game for the war in Afghanistan, in late 2008 a rhetoric of "reconciliation" with the Taliban and other insurgents gained momentum. Practitioners and observers have used the word "reconciliation" to refer to two different kinds of efforts in Afghanistan – lower-level efforts to co-opt the fence-sitters and hired guns, and higher-level negotiations aimed at bringing senior leaders in from the cold.

GEN McKiernan refers to the first group as "small-t taliban", those driven by poverty, lack of jobs and other prospects, and general disaffection, who provide their services to insurgent leaders for some price.[185] The purpose of reconciliation efforts would be to "peel them off" from the hardcore insurgency, perhaps through some combination of economic incentives, opportunities for political participation, and removal from targeting lists. In late 2008, GIRoA was reportedly elaborating a new draft program for small-t taliban reconciliation.[186] To be clear, while GEN McKiernan supports such "reconciliation", he stresses that any such initiatives would be GIRoA – not ISAF or U.S. – efforts.

The second category of reconciliation includes outreach to, and possibly negotiations with, senior leaders of the Taliban and other insurgent groups. The Kingdom of Saudi Arabia has reportedly brokered preliminary talks between representatives of President Karzai – including one of his brothers – and former members of the Taliban who might have access to Mullah Omar and other current senior leaders. President Karzai has stated publicly that efforts to encourage the Saudis to broker contacts have been ongoing for several years, but so far without results.[187] The Saudis have also reportedly facilitated contact with representatives of Gulbuddin Hekmatyar and his HiG organization.[188]

For its part, the Taliban has reportedly named conditions that must be met before it would agree to enter any direct talks. These include the withdrawal of all international forces from Afghanistan, immunity of Taliban leaders from targeting by the ANSF, and the ability to retain their weapons. According to U.S. senior officials, such demands would contradict GIRoA principles – for example, that all Afghan citizens must renounce violence and accept the Constitution – and U.S. Government views.[189] GEN McKiernan has noted that he would have great concern about any measures that would let them remain "the Taliban."[190] For his part, UN SRSG Kai Eide has stressed that the outcome of any talks must be based on the Afghan Constitution – "what is achieved cannot be a compromise," he said, and there should be no "rush to peace."[191]

One further challenge, according to many practitioners and observers, is that despite suffering some tactical-level set-backs, the Taliban leadership appears to feel confident, free to approach any talks from a position of strength. One senior UK official stressed that if negotiations took place today, the Taliban would make unrealistic demands, and he estimated that we are "many months if not years from the end game." He added, "there's no 'quick fix' through reconciliation."[192] Some ISAF officials add that Taliban leaders may be under some pressure from al Qaeda not to participate in negotiations.[193]

Counternarcotics[194]

The Narcotics Problem

One of the things that make the conflict in Afghanistan so intractable is the close linkage between the Afghan insurgency and narcotics. The United Nations Office on Drugs and Crime reported that "warlords, drug lords, and insurgents" collect a tax on the cultivation, transportation and processing of opium poppy – taxes that amounted to "almost $500 million in 2008."[195] The twin counternarcotics (CN) and counterinsurgency challenge is most evident in Helmand province in southern Afghanistan, which is responsible for about two-thirds of Afghan opium production and provides a base of operations for some Taliban insurgents.

Narcotics are big business in Afghanistan – with a climate conducive to cultivation, an absence of readily available economic alternatives such as the infrastructure for bringing legitimate crops to market, and permeable borders with neighboring states. Senior U.S. officials note the reluctance of many Afghan officials to challenge the narcotics industry because its tentacles are so deeply entwined with Afghan governing structures at all levels. Furthermore, Afghanistan still lacks the robust judicial system that would be necessary to prosecute offenders.

Counternarcotics Approaches

The Government of Afghanistan and its international partners have produced no shortage of strategies designed to address Afghanistan's narcotics problem. The current GIRoA plan is the Afghan National Drug Control Strategy, issued in January 2006; a previous version was issued in May 2003.[196] In early 2002, as a follow-on to the Bonn Agreement, the United Kingdom assumed "lead nation" responsibility for coordinating international counternarcotics (CN) efforts; that lead responsibility shifted to GIRoA under the 2006 Afghanistan Compact.

The first premise of the CN effort is the illegality of narcotics. On January 17, 2002, President Karzai issued a decree banning the cultivation, production, abuse, and trafficking of narcotic drugs. Key Afghan CN organizations include: the Afghan Special Narcotics Force (or "Task Force 333"), a paramilitary force created in late 2003 to conduct raids; the Counternarcotics Police of Afghanistan, which investigates and helps target networks; an eradication force that physically carries out eradication; and a CN Criminal Justice Task Force created in early 2005 to expedite CN cases through the fledgling criminal justice system. GIRoA has sought the close cooperation of provincial Governors in its CN efforts. "Governor-led eradication" (GLE) efforts encourage Governors to cooperate by offering them control over the eradication process, and providing them rewards in the form of resources from the central government. Critics have suggested that GLE efforts may selectively target political rivals or small-scale cultivators, while avoiding confrontation with powerful larger-scale producers. In 2006, GIRoA launched the Good Performers Initiative to reward provinces for reducing poppy cultivation.

To support the Afghan CN strategy, and also in an attempt to shape it, the U.S. Government issued its own U.S. Counternarcotics Strategy for Afghanistan, in August 2007.

While GIRoA has tended to favor negotiated eradication – including its collaboration with provincial Governors – the U.S. Government has supported the more forceful approach of centralized and enforced GIRoA-led "non-negotiated forced eradication."[197]

Since assuming command of ISAF, GEN McKiernan has lobbied for a more active NATO role in counternarcotics, including the ability for ISAF forces to target narcotics labs. At an informal meeting of NATO Ministers of Defense, held in Budapest in October 2008, the Allies agreed to stretch the CN role that ISAF can play without changing its formal mandate. After the meeting, NATO Secretary-General Jaap de Hoop Scheffer announced that "under the existing operational plan, ISAF can act in concert with the Afghans against facilities and facilitators supporting the insurgency, in the context of counternarcotics, subject to authorization of respective nations."[198] The following month, GEN McKiernan explained that there was no change to the rules of engagement – just a decision "to be more aggressive." He added, "Where I can make the connection between narcotics, personalities, or facilities, and the insurgency, then I can treat that as a military objective."[199] In December 2008, a senior ISAF official noted that GIRoA had already asked for ISAF's assistance several times, on the basis of October policy update.[200]

Counternarcotics Results

In terms of results to date, according to the United Nations, the Afghan eradication force eliminated 210 hectares of poppy in 2005, 2,250 hectares in 2006, 3,149 hectares in 2007, and 1,174 hectares in 2008. UN data indicates that through Governor-led eradication, 4,000 hectares were eradicated in 2005; 13,050 hectares in 2006; 15,898 hectares in 2007; and 4,306 hectares in 2008.[201] UN data also indicated that the extent of cultivation, and the "farm gate" (preprocessing) value of opium cultivation, both declined in 2008.[202] The U.S. Government reported that in 2008, 18 provinces earned a total of $28 million through the Good Performers Initiative, including $10 million for Nangarhar province, now considered "poppy-free".[203]

Practitioners and outside experts differ concerning how best to evaluate these results. Some simply point to poppy-free provinces – in particular Nangarhar in eastern Afghanistan along the Pakistani border – as successes, and credit a combination of various GIRoA eradication efforts. They suggest in particular the importance of local authorities in discouraging poppy planting. Other observers argue that global food prices may be a more important explanatory variable – in 2008, opium prices were down, while wheat prices, due to a widespread drought, were up, possibly affecting the decision calculus of many planters. They underscore the importance of cultivation trends over time, rather than one-time developments, as better indicators of program effectiveness. Two analysts from this school of thought have argued: "Sustainable reductions in opium poppy cultivation will only be achieved by a wider process of improved security, economic growth and governance."[204]

Capacity-Building as Part of COIN in Afghanistan

According to civilian and military practitioners, the three pillars of the counterinsurgency effort in Afghanistan – security, governance, and development – are inseparably linked. As former advisor to both GIRoA and ISAF Clare Lockhart asserts, "A country is not stable until it has a functioning state that performs key functions for its citizens."[205] International military forces in Afghanistan lead the security line of operation, and also play strong supporting roles in the other two fields, governance and development, particularly at the provincial and local levels where their footprint is much greater than that of civilian counterparts.[206]

The Need for Capacity-Building

Practitioners and observers generally agree that improving the capacity of Afghan institutions is essential for making progress in all three lines of operation – security, governance and development. Many further stress that a critical component of capacity-building in Afghanistan is connecting the center to the regions. Popular support – active or passive – is essential to counterinsurgency, and local-level institutions, whether political officials or the ANSF, are often the most readily-available "face" of government. Local capacity, it is argued, should be competent, and should help connect the local population to a larger "Afghanistan".

One fundamental challenge to capacity-building in general is that Afghanistan's would-be work force was decimated by years of violence and repression, and generally lacks the skills, the professionalism, and often the literacy, to work in a post-Taliban polity or economy.

Some observers argue that the human capital problem was exacerbated by bad choices by the international community in the immediate aftermath of Taliban rule. Eager to place responsibility for leading Afghanistan in Afghan hands – and eager in some cases to focus primarily on the counter-terrorist mission – the international community, it is charged, failed to insist on high standards of experience or integrity in the selection of Afghan leaders at all levels. One ISAF official argued in late 2008, "We need a major housecleaning of GIRoA."[207]

Other observers have stressed that the international community, and particularly the U.S. Government, further aggravates the human capital problem by supporting specific individuals, rather than impartially supporting the Afghan political process as a whole. Such approaches, it is argued, make it difficult to hold such officials accountable. Noting another ramification, NATO SCR Gentilini described such approaches by the U.S. Government as "good but colonialist", because the U.S.-supported officials may be seen as U.S. pawns.[208]

A significant challenge to local-level capacity-building is that provincial Governors and district Administrators have very little formal authority, and they receive no budgets of their own from the central government. As one diplomat described it, in the absence of resources, Governors have to negotiate their authority with *de facto* local power-brokers, which compromises their efforts.[209]

Capacity-Building Initiatives by the Afghan Government

The Government of Afghanistan has launched initiatives of its own aimed at building capacity, particularly at the local level.

In August 2007, GIRoA established the Social Outreach Program (ASOP), under the leadership of the GIRoA-sponsored Independent Directorate of Local Governance (IDLG), led by Director-General Ghulam Jelani Popal.[210] The ASOP is an attempt to harness the power of traditional community organizations to address some local-level governance needs, and to build linkages from the district up to the provincial and national levels. It aims to create temporary community councils at the district level, including between 30 and 40 community leaders, local officials, religious scholars, and tribal heads. The councils are intended to "channel" public needs and grievances, and to provide a forum for conflict resolution, with national-level oversight provided by the IDLG and ministries. District council elections, scheduled for 2010, are expected to replace the temporary councils.

One fundamental challenge is that GIRoA has budgeted no funding to support the effort. As of December 2008, the United States, the United Kingdom, and Canada had agreed to support the program in specific provinces, and talks were underway with other potential sponsors.

In related efforts, in 2008 the IDLG sponsored a series of regional "Governors Meetings", each bringing together a group of provincial Governors, GIRoA ministry officials, and senior officials from ISAF, UNAMA, and Embassies with forces deployed in those provinces. According to some participants, the initial meetings, if they did not produce significant concrete results, underscored the linkages among security, governance, and development, and helped build bridges between the center and the regions.[211]

UN SRSG Kai Eide, in an extension of his role as co-chair of the JCMB, responsible for overseeing implementation of the Afghanistan Compact, actively supports the ASOP initiative.

He has said that, of UNAMA's various efforts, he is particularly proud of this program, designed to draw communities closer to the government.[212]

An earlier GIRoA initiative, the National Solidarity Program (NSP), had a somewhat similar focus. The NSP created community development councils (CDC) that received some money from the central government to spend on priorities they identified. ASOP recognizes this earlier work, particularly the success of some CDCs in identifying local needs, but argues that the CDCs did "not always have the same support and authority within communities as that enjoyed by traditional shuras." ASOP aims to build stronger credibility by basing its efforts more closely on traditional local authority structures.[213]

In addition, some international practitioners have noted that while the CDCs provided some opportunities for local-level innovation, funding streams were inconsistent and then gradually dried up, and some of the councils were "perfectly corrupt."[214]

Military Role and Perspectives on Capacity-Building

ISAF and U.S. military officials in Afghanistan acknowledge that their role in civilian capacity-building is a "supporting" one, but they stress the importance of their role in helping link together the various levels of government. With a footprint that extends through most of the country, and that includes a presence at the local as well as the national and regional levels, the military is well-placed to make such contributions.

The mission statement of the 101st Airborne Division, the nucleus of ISAF's RC-East, underscored the military's role in all three pillars of efforts. It stated that RC-East, in conjunction with GIRoA, ISAF and U.S. civilian agencies, "conducts full spectrum operations to develop Afghan national capability to secure its people, exercise capable governance, and develop a sustainable economy, while defeating terrorists and insurgents, in order to extend GIRoA authority and influence as the legitimate government of the Afghan people."[215]

In 2008, a former U.S. brigade commander in eastern Afghanistan noted that his company commanders worked regularly with district-level officials, monitoring their efforts, and that this focus was important because "the cause of instability in Afghanistan is poor governance."[216]

In late 2008, a senior ISAF official explained that the military's role is to "facilitate" governance, which is "the long pole in the tent," and harder than either security or development. He added that squad and platoon leaders on the ground "regularly liaise with local Afghan officials." Their guidance is to work "bottom-up," to get information, to facilitate shuras, to connect district officials with representatives of Kabul-based ministries…and to follow up assiduously.[217]

Provincial Reconstruction Teams

Provincial Reconstructions Teams (PRT) in Afghanistan grew out of a U.S. military initiative in late 2002. In general, PRTs help Afghan provincial governments develop the capacity and capabilities to govern, provide security, ensure the rule of law, promote development, and meet the needs of the population.[218] The U.S.-led PRT in Zabul province, for example, succinctly states that its mandate is "to conduct civil-military operations in Zabul to extend the reach and legitimacy of GIRoA."[219] As ISAF's area of responsibility expanded geographically, it assumed responsibility for PRTs in each new area. As of early 2009, ISAF maintained 26 PRTs, each led by a single nation. PRT staff may include any combination of civilian and military personnel; the military components formally report to ISAF.

The U.S. Government leads 12 of ISAF's 26 PRTs, 10 of them in RC-East, one in RC-South, and one in RC-West. PRTs do not currently cover all of Afghanistan's 34 provinces. In late 2008, some U.S. officials were considering possible future expansions, including creating two separate PRTs from the single entity currently responsible for Kapisa and Parwan provinces; establishing a PRT-like entity for Kabul city; and creating new PRTs in Dai Kundi and Nimroz provinces in RC-South.

U.S. PRTs are primarily military organizations, each led by a military officer – either an Air Force Lieutenant Colonel or a Navy Commander – who reports to the nearest U.S. battlespace owner. Typically, a PRT includes between 80 and 150 staff members, including one representative each from the Department of State, the Agency for International Development (AID), and the Department of Agriculture (USDA).

Practitioners and observers variously evaluate the successes of PRTs to date. Some argue that while PRTs have carried out useful work, they have not been resourced sufficiently to meet requirements. This may be particularly true for some Allies, for example Lithuania, that have fewer resources available in general for international assistance efforts.[220]

Others, including senior GIRoA officials, have argued that PRTs do not coordinate their efforts sufficiently with Afghan authorities. In November 2008, during a visit to Kabul by a U.N. Security Council delegation, President Karzai claimed that PRTs were setting up "parallel governments" in the countryside.[221] Other GIRoA officials reportedly express that many international resources channeled through PRTs are effectively "lost" amidst multiple layers of contractors and subcontractors, before they reach the Afghan people.[222]

Other U.S. Civilian Field Presence

In addition to the U.S. Government civilian agency presence at PRTs, civilian representatives from the State Department, AID, and USDA also serve at U.S. "Task Forces" – division and brigade headquarters under ISAF. Typically, these civilians provide advisory support, in their respective fields of expertise, helping inform military decision-making and operations. U.S. commanders typically express strong enthusiasm for these partnerships.[223] In

November 2008, a total of 60 U.S. civilians were serving in "field positions" – at PRTs and Task Forces.[224]

U.S. civilian experts work with battlespace-owning military units in another capacity, on Human Terrain Teams (HTT), established in Afghanistan in February 2007.[225] HTTs, recruited and employed by the U.S. Army, are small teams of social scientists, from various academic disciplines, who conduct deep anthropological fieldwork in order to understand and "map" demographic, social and economic dynamics. Their analysis, like the advice of the State, AID and USDA civilian advisors, helps inform military decision-making and operations. U.S. commanders praise the work of the HTTs as contributing directly to their understanding of the battlespace and some have noted that more HTTs – as much as one per district – would be welcome.[226] Their work, like that of servicemembers, entails risk – one member of the HTT in Khowst province, Afghanistan, was killed in May 2008 when his vehicle was struck by an improvised explosive device (IED).

At the same time, the deployment of HTTs has met with both criticism and complications. HTTs have been criticized, most frequently by the academic community, for "compromising" social science ethical standards.[227] In November 2008, an HTT member, Don Ayala, was charged with murder in connection with the killing of an Afghan man who had set a female HTT member on fire, in Kandahar province.[228]

Commanders Emergency Response Program (CERP)

Particularly in the absence of dedicated provincial or district budgets, U.S. commanders have made use of the Commanders Emergency Response Program (CERP), sponsored by the Department of Defense, to fund governance-and development-related projects. One U.S. commander recently called CERP "a surrogate for the government's failure to provide," and another stated, "If we didn't have CERP, we wouldn't be able to do anything."[229]

Since its inception, CERP has provided relatively unconstrained discretionary funds to military commanders on the ground, to meet relatively near-term needs. With far fewer U.S. forces – and thus fewer senior U.S. commanders entitled to spend CERP funds – Afghanistan has received less CERP funding to date than Iraq.

According to U.S. civilian and military officials, and some Afghan provincial officials, decisions about CERP funding allocation are typically based on Afghan priorities and informed by both U.S. civilian and military expertise. The top expenditure to date, by a significant margin, has been road-construction, viewed by Afghans and U.S. officials as critical to security, governance and development.

CERP appropriations for Afghanistan have increased, though not steadily, over time, from $39.7 million in 2004, to $135 million in 2005, to $215 million in 2006, to $206 million in 2007, to $479.9 million in 2008.[230] U.S.-led RC-East noted that $690 million had been requested for Fiscal Year 2009.[231] As additional U.S. forces flow into both eastern and southern Afghanistan in 2009, including a number of additional command positions entitled to spend CERP money, Department of Defense requests for CERP funding may increase.

Agribusiness Development Teams

Supplementing the work of battlespace owners, their civilian advisors, and PRTs in eastern Afghanistan are Agribusiness Development Teams (ADT). ADTs are state-based Army National Guard (ARNG) teams that include "farmer soldiers" who have backgrounds in various facets of agribusiness. The ADTs draw on several decades of similar ARNG experience in Central America, and typically they leverage agricultural expertise from land grant universities in their home states. The teams include organic enablers that allow them to operate independently, including vehicles and force protection.

In 2008, the first ADT, from Missouri, deployed to newly poppy-free Nangarhar province, and the second ADT, from Texas, deployed to Ghazni province. Guardsmen and women from Indiana, Kansas, Kentucky, Nebraska, Oklahoma, and Tennessee have also been involved in the program.[232]

A Civilian Surge?

Many practitioners and observers have suggested that the capacity-building challenges in Afghanistan may require additional international civilian expertise, as well as the effective integration of such expertise with military efforts.

ISAF commanders argue that a stronger commitment to build capacity is required, because it is governance, more than security or development, that is lagging in Afghanistan. RC-East Commanding General MG Schloesser has argued, "We need an interagency surge!"[233] Senior officials from other Allies within ISAF echo this argument – in November 2008, RC-North Commanding General, German Major General Weigt, argued that he needed "civilian advisory teams", as complements to the OMLTs and ETTs. "The main problem for me," he stated," is not security, but deficits in governance."[234]

Outside observers have also argued that the civilian capacity-building effort should be as robust as security capacity-building initiatives. State-building expert Sarah Chayes wrote in December 2008 that the problem of governance in Afghanistan is particularly acute. She argued, "Western governments should send experienced former mayors, district commissioners and water and health department officials to mentor Afghans in those roles."[235]

Some observers have suggested that Afghanistan might be a useful test case for an integrated, balanced application of all instruments of U.S. national power.[236]

In late 2008, U.S. Embassy Kabul outlined a proposal for a "civilian surge" to support provincial-and local-level capacity-building in Afghanistan. The effort would expand the U.S. Government civilian field presence at U.S.-led PRTs and Task Forces, by more than 200 billets. Experts would come from the State Department, AID, and USDA. The additional personnel would augment existing civilian staff functions, and might also establish a presence at the district level to help mentor sub-provincial-level GIRoA officials. In addition, U.S. Embassy Kabul proposed, civilian officials might be given their own "ready funds", to support projects in the field of governance. U S. officials noted that the availability of resources might prove a constraint on the implementation of these proposals, and that even if approved, there was likely to be a time lag before implementation could begin.[237]

FUTURE OPTIONS FOR THE WAR IN AFGHANISTAN

This section outlines and assesses a number of key strategic and operational options for Afghanistan that the U.S. Government, including the U.S. Congress in its oversight capacities, may choose to consider.

Integrating the Overall Approach: Strategy and Implementation

Most Afghanistan observers have pointed to weaknesses in the coordination of the many disparate efforts to support Afghan security, governance or development, including ensuring that those three lines of operation complement one another. Some observers note that leaders of the primary international assistance efforts rarely speak with a unified voice, although their influence might be much stronger were they to do so. The scale of Afghanistan's needs, and the number and variety of entities offering help, make coordination a particularly great challenge.

To address the coordination challenge, some observers advocate crafting a new, overarching strategy for Afghanistan, which would state objectives for security, governance, and development; specify approaches; and assign roles and responsibilities for implementation. A new single set of guidelines, it is argued, would help focus cooperation and coordination.

Many practitioners and observers argue, however, that the basic contours of strategy are already in place. Importantly, GIRoA officials tend to agree with this view. The Afghanistan Compact, an agreement between the Government of Afghanistan (GIRoA) and the international community, identified basic objectives in each of the lines of operation. The 2008 Afghanistan National Development Strategy articulated a strategic vision and key objectives in security, governance, and development; provided multiple individual sector strategies under each of those headings; and stated guidelines for coordinating and monitoring implementation.[238] In the security arena, ISAF's 2008 Joint Campaign Plan (JCP) clearly stated objectives and approaches.

Experts have suggested two alternatives to writing a new strategy, in both of which the U.S. Government could play a leadership role. One school of thought suggests that the specific language of a strategy document may be less important than the "buy-in" to its principles and approaches by all relevant players. In the case of Afghanistan, as former head of UNAMA Lakhdar Brahimi argued, those actors would include all of Afghanistan's neighbors, other regional leaders "including India, Iran, and Saudi Arabia," all permanent members of the UN Security Council, and all major donors including Japan.[239] What is needed, this school suggests, are intensified diplomatic efforts to forge a broad consensus on the basic contours of existing strategy.

Other experts have suggested a second option – accepting the basic tenets of existing strategy documents and focusing on strengthening monitoring and implementation. Diplomatic efforts might urge contributing countries to more rigorously tailor their assistance efforts to Afghan national priorities, or they might attempt to curb the "national-first" approach of those contributing countries that regularly lobby GIRoA ministers for attention to "their" provinces. New efforts might also support strengthening UNAMA so that it might more effectively play its critical coordination role as co-chair of the Joint Coordination and Monitoring Board (JCMB), tasked to oversee coordinate implementation of the Afghanistan Compact.

Updating U.S. Grand Strategy for Afghanistan

As a leading protagonist of international support for Afghanistan, the U.S. Government has played a strong role in shaping the Afghanistan Compact, NATO's strategic vision, and ISAF's Joint Campaign Plan, all of which articulate strategic objectives and approaches for Afghanistan. Nevertheless, as a normal manifestation of international relations, specific U.S. national interests in Afghanistan and the region, and prioritized U.S. objectives designed to support those interests, may differ somewhat from broadly agreed international goals. Therefore, some experts suggest the utility of a rigorous internal review of strategic objectives and approaches, and a new articulation of national strategy, by the U.S. Government.

As of early 2009, several such reviews were underway or had recently been completed, including a review of strategy for Afghanistan and Pakistan by the George W. Bush White House, a study by the Joint Staff, and a 100-day strategic assessment of its entire area of responsibility by CENTCOM. Options for the Obama Administration might include publishing a formal, public, unclassified Afghanistan strategy document, or issuing internal guidance concerning the premises of U.S. strategy for Afghanistan for U.S. Government practitioners. Congress has required reports articulating Administration strategy in the past, and could do so again.

Many experts suggest that U.S. Government strategic reviews and conclusions would benefit from a sober assessment of Afghanistan's starting point, including its poverty, dearth of human capital, lack of resources and development, and lack of tradition and experience as a functioning, effective government. That starting point, it is argued, might usefully help temper the expectations and assumptions underlying U.S. strategy.

Adjusting U.S. and Other International Forces

In 2008, NATO and U.S. commanders on the ground in Afghanistan requested additional international forces – requests echoed by GIRoA officials. Key policy questions include the scope of the requirement, the likely duration of the requirement, and the options, including possible combinations of U.S. and other NATO forces, for meeting the requirement.

For those who advocate troop level increases, the duration of the need for higher international troop levels in Afghanistan is hard to determine in advance. Future requirements might depend on changes in the strength of the insurgency, growth in the capacity and capabilities of the ANSF, adaptations in the approaches used by international forces, and concurrent progress in the areas of governance and development. GEN McKiernan has argued that international troop levels need to be increased until the "tipping point" is reached when the ANSF assume lead security responsibility – "three or four more years away."[240] To be clear, Afghan and U.S. officials do not use the term "surge", arguing that "surge" refers to a temporary increase in troop strength, while the requirement in Afghanistan is likely to last for some time.[241]

Requirement for Additional Forces?

A first premise of counterinsurgency is conducting operations "by, with and through" indigenous forces. Where such forces are not available, international forces may substitute, support, and/or help build additional indigenous forces. The military effort in Afghanistan was, from the start, an "economy of force" mission, in terms of both international troop strength, and early target endstrength goals for the Afghan National Security Forces (ANSF). That approach provided only limited geographical coverage of Afghanistan. Commanders on the ground argue that significant additional forces are needed to meet the full scope of requirements.

At the same time, some key observers caution against additional U.S. or other international force deployments, on the grounds of general Afghan antipathy to the presence of foreign forces on their soil, exacerbated by any episodes of heavy-handedness by those forces. Thus, they suggest additional deployments could actually prove counterproductive to the COIN effort. Regional expert Rory Stewart has argued flatly, "the West should not increase troop numbers," because doing so would inflame Afghan nationalism and lend support to the insurgency. "The Taliban," he adds, "which was a largely discredited and backward movement, gains support by portraying itself as fighting for Islam and Afghanistan against a foreign military occupation."[242] Noted regional scholar Barnett Rubin has argued, similarly, that the "Afghans don't like their country being occupied by foreign soldiers any more than did their ancestors," and that reaching a political solution to the insurgency "may require decreasing the U.S. and other foreign military presence rather than increasing it."[243]

NATO Forces

One option for meeting any identified requirements for additional forces is through contributions from NATO Allies. Secretary of State Clinton has indicated the intention of the Obama Administration to use the opportunity of NATO's 60th anniversary, in April 2009, to seek additional contributions from Allies.[244] Key considerations include both the likelihood and the utility of additional NATO contributions.

Some experts have raised questions about the utility of possible contributions from some NATO Allies, given the national caveats that tightly constrain the activities of many contingents. One option for the U.S. Government would be to continue to press Allies to relax or eliminate such caveats – Secretary of State Clinton has indicated that the Obama Administration intends to pursue this approach.[245] One alternative in some cases would be to focus on contributions less likely to create domestic political opposition. Some contributors, for example, might provide so-called "niche" capabilities such as explosive ordnance disposal (EOD) expertise and medical facilities, while others might choose to "sponsor" – that is, pay for and run – branch schools used for training the Afghan National Army.

U.S. Forces

The most likely avenue for meeting any requirements for additional international forces is an increase in U.S. troop strength in Afghanistan. As of the beginning of 2009, an additional brigade combat team (BCT) – 3rd BCT, 10th Mountain Division – was flowing into Regional Command-East, and GEN McKiernan had requested three additional BCTs or equivalents, plus an aviation brigade and all the necessary enablers. The Combined Security Transition Command-Afghanistan (CSTC-A), in turn, had repeatedly stressed a requirement for additional trainers – 2,300 for the Afghan police, and 60 teams for the ANA. Key questions include how much U.S. force presence is enough, what balance of forces is needed, and for how long.

One factor that might have an impact on required U.S. troop numbers, and how those requirements are described, is force employment. GEN McKiernan notes that the plan for many additional U.S. forces expected to flow into Afghanistan in 2009 is "dual-missioning," in which a given unit owns geographical battlespace and also plays a training and advisory role with the ANSF. For example, a BCT might be "plussed up" with additional Military Police (MP) structure, to help advise the Afghan National Police (ANP).[246] The U.S. Marine Corps Task Force that includes 3rd Battalion, 8th Marine Regiment, that arrived to serve in southern Afghanistan in late 2008, performs such a dual mission. Such arrangements might affect the ways that requirements are described – deploying a BCT "plussed up" with additional MPs, for example, might meet part of the requirements for both combat brigades and police trainers.

Observers have raised several different concerns about the ramifications of substantially increasing the U.S. share of the total ISAF force, as well as strengthening the U.S. role in ISAF command and control arrangements. The more prominent the U.S. role, some argue, the less likely that Allies will increase or perhaps even sustain their own contributions, since it will be easier for them to argue that the United States is meeting the requirement. Other skeptics point to a potential operational challenge – the possibility of a *de facto* bifurcation of Afghanistan into two distinct sets of approaches, robust COIN by the United States and a few Allies in eastern and southern Afghanistan, and softer stability operations by all other Allies and partners in northern and western Afghanistan. Such visible disparity, if it emerged, would make it much harder for ISAF to speak with a unified voice to GIRoA. Many civilian and military senior ISAF officials from non-U.S. Allies gently caution that, at the very least, the United States should avoid suggesting that the increased U.S. role was made necessary by inadequacies in the performance of their Allies.

Logistics

The deployment of additional U.S. forces is likely to raise significant logistics challenges. U.S. forces based in eastern and southern Afghanistan have relied heavily to date on lines of communication (LOC) running across Afghanistan's mountainous eastern border into Pakistan, and down to the port city of Karachi. According to ISAF officials, with increased force flow, the demand on those LOCs is likely to increase by 500%, posing challenges in terms of both capacity and security. In late 2008, a series of insurgent attacks on those LOCs, on the Pakistani side of the border, underscored their vulnerability. In addition, the Government of Pakistan (GoP) has occasionally closed the primary border crossing, at the Khyber Pass, ostensibly to support GoP military operations in border regions. "Our biggest vulnerability," said one ISAF senior official, "is our LOCs."[247]

One alternative is the use of northern supply routes through former Soviet republics north of Afghanistan. On January 20, 2009, GEN Petraeus, CENTCOM Commanding General, announced that new agreements had been reached with Russia and Central Asian states regarding the transit of goods and supplies to Afghanistan. These arrangements are intended to complement the Pakistani supply routes – GEN Petraeus commented, "It is very important as we increase the effort in Afghanistan that we have multiple routes that go into the country."[248] Some experts have raised a concern about this option – the uncertainty about Russia's future orientation, its support for the international effort in Afghanistan, and the influence it might choose to exercise over any Central Asian support to that effort.

One additional practical consideration is the additional requirement for contractors resulting from increased U.S. troop deployments. ISAF officials noted in late 2008 that the U.S. Army had insufficient engineering assets available to support the construction in southern Afghanistan required by the arrival of large numbers of additional U.S. troops. They added that the availability of local contractors in southern Afghanistan was already scarce, given that the limited pool of qualified personnel was already occupied with work for ISAF and other organizations.

Sourcing U.S. Deployments

With concurrent wars in Afghanistan and Iraq, the Department of Defense (DoD) could find it a challenge to source and then sustain larger troop deployments to Afghanistan.

One ramification could be delays in the arrival of additional forces in theater. Department of Defense (DoD) leaders indicated in late 2008 that DoD was working to send additional military forces to Afghanistan but that it would take some time – perhaps half a year – for them to become available. Senior U.S. military officials have indicated that additional force flow into Afghanistan is connected to U.S. force drawdowns in Iraq. Some observers note that such a delay may not prove neutral – that even if the insurgency is held to what some practitioners call its current "stalemate", Afghan popular opinion, the center of gravity in COIN, may well slide toward greater disaffection.

A second implication could be a continuation of stress on parts of the force. Numerous defense strategists have commented on the stress that simultaneous wars in Afghanistan and Iraq have placed on U.S. military forces, and particularly on the Army and Marine Corps. That pressure has manifested itself in repeated and extended deployments for both units and individual servicemembers, and has affected the personal lives of individual servicemembers through medical conditions and stress on families.

One key decision concerning additional U.S. force flow to Afghanistan will be the mix of Military Services and in particular, the Army and Marine Corps. The Marine Corps has long advocated a transition of its primary focus from Iraq to Afghanistan, arguing that improved security conditions in Anbar province, its area of responsibility in Iraq, allow a drawdown, and that Marine warfighting capabilities are especially well-suited for the growing security challenges in southern Afghanistan.[249]

A further decision concerns the type of Army combat forces that would deploy. To date, the U.S. Army has deployed "light" combat units to Afghanistan. If the Army continues to provide the majority of U.S. forces in Afghanistan, and if the decision is to continue to deploy exclusively light combat units, then stress on the infantry could become acute. One option might be the introduction of heavy BCTs, but their capabilities might be less well-suited for Afghanistan's terrain.

Updating Counterinsurgency (COIN) Approaches

Counterinsurgency approaches, by definition, require continual adaptation to the local environment.

One set of COIN military strategy questions in Afghanistan concerns population security, and in particular, how best to balance the need to maintain a dispersed footprint, living among the population, given the antipathy of some parts of that population toward any foreign presence. In Afghanistan, this dilemma has been particularly evident in some isolated mountain valleys. One option is to sustain an international troop presence, coupled with development and governance assistance, aiming to win over the local population eventually. An alternative – diametrically opposed both theoretically and practically – is to accept that "population security" may not work identically in every context, to withdraw the antagonizing foreign force presence in some locations, and to sacrifice some measure of security against insurgents for a different kind of stability – an absence of discord between foreign forces and the local population.

Another set of options concerns possible adjustments to the use of kinetic air operations in Afghanistan. The year 2008 witnessed sharp criticism from GIRoA and UNAMA of civilian casualties caused by air strikes. Such incidents have the potential to sharply discourage popular support for the government, essential to the success of any counterinsurgency. In late 2008, ISAF updated its Tactical Directive, carefully articulating rules for the use of fires. The year 2009 may prove a good test of the impact on civilian casualties of the 2008 Directive, and of a larger international forces presence that may allow some reduction in the use of kinetic air power.

A further set of options concerns the balance of effort between Special Operations Forces (SOF) and conventional forces. Some experts have argued that the military effort in Afghanistan would benefit from boosting the role of SOF vis-à-vis conventional forces. In Afghanistan, SOF play the lead role in targeting insurgent leaders, and proponents argue that increasing SOF troop strength might lead to more targeting successes. Other experts counter that successful SOF efforts rely on international conventional forces and the ANSF, both to provide information about a battlespace in advance, and to "hold" that battlespace after kinetic operations. That collaborative partnership, they argue, is essential to the success of overall COIN efforts, and therefore, the utility of unilateral increases in SOF deployments for the overall COIN mission would be limited.

Developing the Afghan National Security Forces (ANSF)

Experts generally agree that the development of the Afghan National Security Forces is essential to the future security and stability of Afghanistan.[250] In late 2008, GIRoA, backed by the support of the international community, decided to increase the target endstrength of the Afghan National Army (ANA). Further major policy issues include the size of the total ANSF, the mix of forces within it, the focus of effort of those forces, the timeline for developing those forces, and funding further ANSF development.

Practitioners underscore that ANSF development is highly dynamic. As Afghan National Police (ANP) capabilities grow, the ANP may increasingly assume responsibility for some domestic missions now performed by the Afghan National Army (ANA). As total ANSF numbers and capabilities grow, Afghan forces may increasingly assume responsibility from international security forces. And when – as expected – the security challenges from the insurgency diminish, the ANSF may shift from COIN to more traditional peacetime foci, including external defense for the ANA, and civilian law enforcement for the ANP.

One critical issue is funding the development and sustainment of the ANSF. Senior Afghan and international officials estimate that it will cost approximately $3.5 billion per year to increase ANSF force structure, and then $2.2 billion per year to sustain it. Unlike Iraq, whose oil revenues have funded an increasing share of the costs of growing and sustaining the Iraqi Security Forces in recent years, Afghanistan has few natural resources and little economic activity, other than poppy production, that could generate significant revenue in the near future. GIRoA, which contributed $320 million to the ANSF in 2008, is not a realistic source of ANSF funding in the near term.[251] International support, and particularly U.S. support, is expected to bear the near-term burden of developing the ANSF, until it reaches its current endstrength targets.

Growing the ANA to 134,000 – or more – raises the twin questions of funding and sustainability. It is expected that the currently planned ANA growth will be funded by the international community; the United States is currently the leading contributor. If GIRoA wanted to sustain the force beyond that time frame, theoretical options would include continued U.S. and international support, or Afghan assumption of some level of financial responsibility. Alternatives could include demobilizing some part of the force – if GIRoA had the ability to do so – or, hypothetically, making part of the force available to serve in multi-lateral peace operations, in which case the international community might bear some of its costs.

For the future, one option, in the absence of GIRoA ability to shoulder the burden, would be sustained international support, a responsibility likely to fall to the U.S. Government, based on current patterns. The policy question for the U.S. Government, in turn, would concern whether national interests support sustaining that sizable commitment – and if so, at what levels, and for how long.

In the absence of either Afghan or international funding sources, one option would be demobilizing some part of the Afghan forces. A conclusive defeat of insurgent forces might increase the plausibility of such an approach, by reducing the requirement for ANSF numbers for the COIN fight. One challenge could prove to be limited ability on the part of the Afghan civilian leadership to direct the security ministries to decrease their forces.

One additional option, with some cost implications, would be a future role for Afghanistan as net exporter of security, providing trained, interoperable ANSF to serve in UN peace operations or other multi-national coalition efforts around the world. Minister of Defense Wardak suggested this possibility by noting, “One day we will pay our debt by fighting shoulder to shoulder with you,” words that might apply abroad as well as at home.[252] Playing such roles might help GIRoA secure some international support for maintaining its force.

A further policy issue concerning the ANSF is force modernization – including the likely future requirements the ANSF will face, and the steps the ANSF might begin taking now, in terms of training and procurement, to prepare to meet those future requirements. While army and police commanders on the ground are fully focused on the current fight, Minister of Defense Wardak has argued in favor of beginning to build a mix of forces that would also be appropriate to the post-COIN environment. One implication for the U.S. Government is the impact GIRoA force modernization decisions may have on foreign military sales requests.

One final consideration concerning the development of the ANSF, is the nature of emerging civil-military relationships within the Government of Afghanistan. Some civil-military experts caution that there is an inherent danger, when a state’s army is by far its most competent, effective organization, that civilian control of the military may erode. In late 2008, some Afghan officials, including Minister of Defense Wardak, have pointed to a tendency on the part of President Karzai to ask the ANA to play non-military roles – for example, guarding civilian prisons – when civilian personnel are unavailable. U.S. military officials have noted that they spend significant time with senior Afghan security officials, mentoring them on the role of security forces in a democracy.[253]

Addressing Regional Issues

By the end of 2008, a near-consensus had emerged among practitioners and observers that the challenges Afghanistan faces could not be successfully addressed without a "regional" approach. Policy options might include various avenues for influencing the behavior of neighboring states, particularly Pakistan and Iran, in order to shape the security climate in Afghanistan. Some but not all commentators argue, in addition, that because Pakistan and India tend to make use of Afghanistan as a pawn in their own long-simmering conflict, successful "regional" initiatives concerning Afghanistan must also take into account the Pakistan-India relationship. Some commentators also suggest that Iraq is germane as a major security preoccupation for Iran, whose behavior in Afghanistan may be affected by developments across its opposite border. While the exhortation to consider the broader region has widespread support, no consensus on regional strategy has emerged – to date, experts have put forward a very wide range of specific policy prescriptions, some of them mutually contradictory.

Pakistan

The U.S. Government may wish to address both strategic and operational policy options concerning Pakistan's role vis-à-vis Afghanistan.

At the strategic level, a central question concerns the most effective ways to encourage Pakistani action against those supporting and fomenting the Afghanistan insurgency. One group of experts urges assistance efforts to support governance and development in "the impoverished areas that have become breeding grounds for militants", in particular along the border, in order to diminish the generation of, and support for, insurgents.[254]

Several other groups of experts particularly emphasize the importance of the U.S.-Pakistani mil-to-mil relationship, as a tool for shaping Pakistani actions. Some argue that the key is to continue to build Pakistani capabilities and to shape their orientation by strengthening current initiatives – for example, by boosting U.S. efforts to train the Pakistan Army Special Forces for counter-terrorism operations, expanding Pakistani participation in military exchange programs, and fostering closer mil-to-mil coordination along the Afghanistan-Pakistan border.[255] Others agree that mil-to-mil initiatives are key but urge making the provision of the benefits they offer conditional on stronger Pakistani support along the border with Afghanistan.[256] Some experts, in contrast, argue that the close U.S.-Pakistan mil-to-mil relationship is part of the problem, as it may tend to diminish Pakistani civilian control over its military. Therefore, one regional expert writes, "the United States must relinquish, not strengthen, the privileged relationship between the United States and the Pakistani military."[257]

Another set of experts argues that efforts to shape Pakistani actions vis-à-vis Afghanistan are likely to be successful only as part of a broader approach to Pakistan's concerns and its role in the region. Key issues, in this broader perspective, might include Pakistan's role as a nuclear power, and its tendentious relationship with nuclear-armed neighbor India – in particular their rival claims to the disputed territory of Kashmir. Policy options might include, for example, more robust U.S. unilateral, or multilateral, diplomatic efforts to help India and Pakistan forge a stable, sustainable relationship.

The Obama Administration has indicated initial support for a Pakistan policy that includes both increasing non-military aid, and conditioning the provision of mil-to-mil assistance on Pakistani actions such as "...clos[ing] down training camps, evict[ing] foreign fighters, and prevent[ing] the Taliban and al Qaeda from using Pakistan as a sanctuary."[258]

Meanwhile, at the operational level, a key question is whether to continue U.S. cross-border attacks while strategic-level dialogue and other attempts to exercise leverage are underway. Continued cross-border attacks might lead to political blow-back in Pakistan, but they might also persuade Pakistani officials to take more decisive steps to end support to insurgents. Options, therefore, might include either accelerating or decreasing the use of cross-border attacks. Options might also include seeking additional avenues for strengthening tactical-level cross-border coordination, including increasing the frequency or scope of regular coordination meetings, establishing further border coordination centers, and improving the ease of tactical-level coordination in actions against insurgent targets.

Iran

Practitioners and observers recognize Iran's dual track role in Afghanistan – providing both humanitarian and lethal aid. The key issue in U.S. Afghanistan policy debates is how best to leverage concerns that Iran shares with the rest of the international community – including regional security and the impact of the narcotics trade – to shape Iranian choices.

Options include a more concerted diplomatic outreach to Tehran, by the United States and/or other members of the international community, designed to build on shared regional concerns. Some experts suggest that Tehran may be more likely to respond positively to outreach efforts by the Obama Administration, than to those by its predecessor. Another option – not mutually exclusive – is robust countering of any Iranian efforts to provide lethal aid to insurgents in Afghanistan.

Strengthening Counternarcotics Efforts

Despite the existence of broadly agreed counternarcotics (CN) strategies, and some recent declines in cultivation, many practitioners see little likelihood of significant further CN progress until GIRoA takes major further strides in establishing the rule of law, including both building a formal judicial system and encouraging a pervasive law-based culture. Some observers have further suggested that there may be inherent contradictions between CN and COIN efforts, since a key premise of COIN is fostering popular support for the government, while some CN initiatives may alienate parts of the population, at least temporarily.

One near term option is for GIRoA and ISAF to take full advantage of the October 2008 NATO decision expanding the interpretation of the CN role ISAF may play, to include targeting drug facilities when a connection with the insurgency can be shown. Focused diplomatic efforts might prove useful in persuading some Allies to fully embrace this broader interpretation.

A further option might be more assiduously cultivating the cooperation of community as well as provincial leaders in CN efforts, including strengthening the system of incentives available to those who lend their support. The Afghanistan Social Outreach Program, designed to foster and focus ground-up capacity-building, might naturally complement any such outreach efforts.

Achieving a NATO Success in Afghanistan?

Most NATO observers suggest that "Afghanistan" is a critical test for the Alliance, including its ability to conduct major out-of-area missions, and its relevance to 21st century security challenges, and many have argued that failure in Afghanistan could spell the end of the Alliance.

The issue of the role of NATO in Afghanistan is likely to gain prominence on the world stage in early 2009, amidst preparations for NATO's 60th anniversary summit, scheduled for April 2009. In January 2008, the Afghanistan Study Group argued, "A failure of the NATO mission in Afghanistan would also damage the future prospects of the organization itself."[259] For a number of practitioners, that line of thinking implies an imperative to make sure that the Alliance is successful in Afghanistan – which is not quite the same thing, logically, as making sure that Afghanistan itself succeeds. Key policy considerations include what it would take for the outcome in Afghanistan to be considered a "NATO success", and what if any differences that might entail from a strict pursuit of security, good governance and development for Afghanistan. In some circumstances, U.S. policy-makers might wish to weigh the imperative of accomplishing the mission against the imperative to support NATO as an institution.

Some observers suggest that the growing presence of U.S. forces, and possibly a stronger U.S. role in ISAF command and control, may lead to the perception that any progress in Afghanistan is due to U.S. rather than NATO efforts. In that vein, the Afghanistan Study Group argued, "Burden-sharing among NATO allies is critical to the mission in terms of both available resources and public perceptions – an increasingly unilateral mission will be politically vulnerable in Afghanistan, the U.S., and NATO."[260] Options for countering such perceptions, if so desired, might include maintaining U.S. battlespace-owning units under ISAF command; using diplomatic channels to forge stronger consensus on the nature and stakes of the fight in Afghanistan; and soliciting non-military forms of assistance from those countries unwilling or unable to provide large ground force contingents for the counterinsurgency fight.

Supporting Reconciliation

A number of key practitioners and observers have supported "reconciliation" outreach initiatives, to insurgent leaders and/or their foot soldiers, as one avenue toward final settlement of the conflict in Afghanistan. Senior U.S. military officials in Afghanistan underscore that questions concerning to whom to reach out, at what time, with what offers, and with what endstate in view, are policy matters for GIRoA decision. Any "reconciliation" efforts would likely have a direct bearing on security efforts in Afghanistan, because they concern the formal adversary – the insurgents – in the fight.

Some experts have argued that leadership-level reconciliation initiatives risk institutionalizing formal political roles in the future Government of Afghanistan for known "bad guys". Such inclusion, they suggest, might alienate parts of the Afghan population who suffered repression under the Taliban regime. Worse, such alienated constituencies might take up arms to protest such brokered deals. Institutionalizing leading roles for former Taliban or other insurgent leaders might also, it is suggested, push the orientation of the Government of Afghanistan in more repressive directions.

U.S. Government policy considerations might include determining the U.S. preferred outcome of any such talks, and exercising diplomacy to influence the form and content of any such initiatives. From an operational perspective, U.S. deliberations might include assessing how different military strategies might contribute to bringing Taliban leaders to the negotiations table in a frame of mind conducive to agreeing to conditions acceptable to GIRoA. For example, Taliban leaders might be compelled to come to terms by even more aggressive targeting, or by more strident efforts to cut off the funding support they receive from the narcotics trade.

Applying the Lessons of Afghanistan to Future Force Development

For the longer term, defense practitioners and analysts are likely to continue to seek lessons from U.S. military prosecution of the wars in Afghanistan and Iraq to apply to future U.S. force shaping and sizing. Such conclusions, and they way they are applied, are likely to have a profound impact on how the Military Services fulfill their responsibilities, in accordance with Title 10, U.S. Code, to organize, man, train and equip military forces. Critical policy considerations for DoD are likely to include the capabilities required to successfully prosecute complex contingency operations like those in Afghanistan and Iraq, the likelihood that the United States will engage in similar complex contingencies in the future, and the relative importance of such skills compared to more traditional military capabilities.

The publication of Department of Defense Directive 3000.07 on "Irregular Warfare", in December 2008, which stated that "IW is as strategically important as traditional warfare," reflected and helped institutionalize a growing DoD emphasis on complex contingency capabilities.[261]

Should this school of thought continue to hold sway, options might include further increasing the endstrength of the ground forces, and raising the profile of the mission to train and advise foreign security forces. In a resource-constrained environment, any such choices would like entail tradeoffs with other DoD capability areas.

Balancing and Integrating Civilian and Military Efforts

Years of operational experience in Afghanistan, like those in Iraq, have helped fuel debates about appropriately balancing military and civilian capabilities in the U.S. Government, and effectively integrating those capabilities with each other.

A key short-term policy consideration is the U.S. Government response to the appeal from U.S. Embassy Kabul for further civilian resources – personnel and ready funds – to support capacity-building efforts in Afghanistan. One constraint may be the availability of trained civilian personnel with the appropriate expertise, ready to deploy. Another constraint may be the availability of funds – in particular, civilian funding streams relatively free of bureaucratic red tape that allow quick execution.

Some U.S. commanders on the ground, while generally welcoming the prospect of civilian expertise, caution that such an initiative would be most valuable if it carefully recruits people with relevant expertise. Simply "throwing bodies at the problem," they caution, would not be helpful. Some add that it would also be important to ensure that additional civilian personnel are included in clear civilian chains of command, to ensure unity of effort among their various activities.

Some outside experts, in turn, caution against any infusion of civilian personnel that might detract from GIRoA efforts to govern their own country. One regional expert, arguing that the U.S. Government should not increase its involvement in the Afghan government or economy, has written: "The more responsibility we take in Afghanistan, the more we undermine the credibility and responsibility of the Afghan government and encourage it to act irresponsibly."[262] Another expert has urged, "Rather than sending in thousands of civilians, the shift in emphasis could be to training Afghans to do the jobs themselves."[263] U.S. Embassy appeals for a "civilian surge" are predicated on precisely that advisory approach.

Key policy considerations for the longer-term, suggested by U.S. experience in Afghanistan, might include whether the U.S. government requires increased civilian capacity to meet possible future complex contingency requirements, and if so, what capabilities would be required, and how might they best be cultivated and organized. A related consideration concerns the implications of any enhancement of U.S. government civilian capacity for military requirements – whether, for example, increased civilian capacity might decrease the missions military forces should be prepared to accomplish. And finally, regardless of the balance in numbers and capabilities between military and civilian personnel, a further policy consideration is how best to effectively integrate military and civilian complex contingency efforts, in both planning and execution.

Additional Reports

CRS Report RL33110, *The Cost of Iraq, Afghanistan, and Other Global War on Terror Operations Since 9/11*, by Amy Belasco.

CRS Report RL32686, *Afghanistan: Narcotics and U.S. Policy*, by Christopher M. Blanchard.

CRS Report RL33627, *NATO in Afghanistan: A Test of the Transatlantic Alliance*, by Vincent Morelli and Paul Belkin.

CRS Report RL30588, *Afghanistan: Post-War Governance, Security, and U.S. Policy*, by Kenneth Katzman.

CRS Report RL33498, *Pakistan-U.S. Relations*, by K. Alan Kronstadt.

Source: The University of Texas at Austin, Perry-Castañeda Library Map Collection, Afghanistan Political Map 2003.

Figure 1. Map of Afghanistan

AUTHOR CONTACT INFORMATION

Catherine Dale
Specialist in International Security
cdale@crs.loc.gov, 7-8983

End Notes

[1] See for example the replies to questions for the record, submitted by Secretary of State nominee Hillary Clinton to the Senate Foreign Relations Committee (SFRC), for her January 13, 2009, confirmation hearing, available at http://www.foreignpolicy.com/files/KerryClintonQFRs.pdf; and replies to questions for the record, submitted by Under Secretary of Defense for Policy nominee Michèle Flournoy to the Senate Armed Services Committee (SASC), for her January 15, 2009, confirmation hearing, available at http://armed-services.senate.gov/statemnt/2009/January/Flournoy%2001-15-09.pdf. See also Joseph J. Collins, "Afghanistan: Faltered But Not Fallen," *Armed Forces Journal*, January 2009; Michael O'Hanlon, "Playing for Keeps," *USA Today*, January 7, 2009; Nathaniel C. Fick and John A. Nagl, "Counterinsurgency Field Manual: Afghanistan Edition," *Foreign Policy*, January/ February 2009.

[2] President Barack Obama, Inaugural Address, January 20, 2009, text available at http://www.whitehouse.gov/blog/read_the_inaugural_address/.

[3] In her replies to questions for the record, submitted to the Senate Foreign Relations Committee (SFRC) for her January 13, 2009, confirmation hearing, Secretary of State nominee Hillary Clinton stated: "President-Elect Obama and I believe that Afghanistan and the Pakistani border are the central front in the war on terror." See text available at http://www.foreignpolicy.com/files/KerryClintonQFRs.pdf.

[4] For example, the *National Defense Authorization Act for Fiscal Year 2008*, P.L. 110-181, January 28, 2008, §1230 required a "report on progress toward security and stability in Afghanistan", no later than 90 days after enactment and every 180 days thereafter until the end of Fiscal Year 2010. That report is to include a "comprehensive strategy of the United States for security and stability in Afghanistan" that addresses NATO and its International Security Assistance Force (ISAF), the Afghan National Security Forces (ANSF), Provincial Reconstruction Teams (PRTs) and other development initiatives, counter-narcotics activities, the rule of law, and regional considerations. P.L. 110-181, §1231, required, no later than 90 days after enactment, and annually to the end of Fiscal Year 2010, a report on sustaining the Afghanistan National Security Forces.

[5] Babak Dehghanpisheh and Evan Thomas, "Scions of the Surge," *Newsweek*, March 24, 2008.

[6] Field Manual 3-24, *Counterinsurgency*, Headquarters, Department of the Army, December 2006, para. 2-1. Paragraph 2-2 of the COIN manual adds: "The integration of civilian and military efforts is crucial to successful COIN operations. All efforts focus on supporting the local populace and HN [host nation] government. Political, social, and economic programs are usually more valuable than conventional military operations in addressing the root causes of conflict and undermining an insurgency."

[7] In November 2008, International Security Assistance Force Commanding General, U.S. Army General David McKiernan noted that without additional resources, it would be "a longer fight with greater sacrifices." GEN David McKiernan, Interview, Kabul, Afghanistan, November 2008.

[8] For background see Steve Coll, *Ghost Wars: The Secret History of the CIA, Afghanistan, and Bin Laden, from the Soviet Invasion to September 10, 2001* (New York: Penguin, 2004); George Crile, *Charlie Wilson's War: The Extraordinary Story of How the Wildest Man in Congress and a Rogue CIA Agent Changed the History of our Times* (New York: Grove Press, 2003); Robert D. Kaplan, *Soldiers of God: With Islamic Warriors in Afghanistan and Pakistan* (New York: Vintage Departures, 2001); and Ahmed Rashid, *Taliban: Militant Islam, Oil and Fundamentalism in Central Asia* (New Haven, CT: Yale University Press, 2001).

[9] The plural noun "mujahedin" (singular "mujahid"), borrowed from Arabic and now used in standard English, refers to a group of Muslims waging "jihad", or "a holy war waged on behalf of Islam as a religious duty." See "jihad", Merriam-Webster Online Dictionary 2008, Merriam-Webster online, available at http://www.merriamwebster.com/dictionary/jihad; and "mujahideen," *Merriam-Webster Online Dictionary 2008*, Merriam-Webster online, available at http://www.merriam webster.com/dictionary/mujahideen>.

[10] The term "Taliban," in Pashto, is the plural of "talib" (student), which is derived from Arabic. See "Taliban," *Merriam-Webster Online Dictionary 2008,* Merriam-Webster online, available at http://www.merriam-webster.com/dictionary/Taliban.

[11] The full list of demands included: "Deliver to United States authorities all of the leaders of Al Qaeda who hide in your land. Release all foreign nationals, including American citizens you have unjustly imprisoned. Protect foreign journalists, diplomats and aid workers in your country. Close immediately and permanently every terrorist training camp in Afghanistan, and hand over every terrorist and every person and their support structure to appropriate authorities. Give the United States full access to terrorist training camps, so we can make sure they are no longer operating." See President George W. Bush, Address to Joint Session of Congress, September 20, 2001, available at http://www.whitehouse.gov/news/releases/2001/09/20010920-8.html.

[12] See Statement by President George W. Bush, October 7, 2001, available at http://www.whitehouse.gov/news/releases/2001/10/20011007-8.html. Many observers consider that at the launch of OEF, short-term U.S. objectives – including targeting al Qaeda – together with the means to achieve them, were much more clearly articulated than any longer-term U.S. vision for Afghanistan's future, together with the approaches necessary to achieve that vision.

[13] The United Kingdom's publicly stated campaign objectives included bringing Osama bin Laden and other al Qaeda leaders to justice; preventing them from posing a further terrorist threat; and ensuring that Afghanistan ceased to harbor terrorists; in pursuit of the broader objective to "do everything possible to eliminate the threat posed by international terrorism." See Ministry of Defence, United Kingdom, "Defeating International Terrorism: Campaign Objectives," October 16, 2001, available at http://www.mod.uk. For a detailed discussion of the March 2002 Operation Anaconda, which included SOF and conventional forces, coalition partners, and Afghan forces, see Sean Naylor, *Not a Good Day to Die: The Untold Story of Operation Anaconda* (New York: Berkley Books, 2005). For an analysis of the lessons of Afghanistan operations for future warfighting, see Stephen Biddle, Afghanistan and the Future of Warfare: Implications for Army and Defense Policy, Carlisle, PA: Strategic Studies Institute, November 2002.

[14] In accordance with the provisions of the Bonn Agreement, a large meeting – a "loya jirga" – was held in June 2002, at which Hamid Karzai was elected head of the new Afghan Transitional Authority. A new constitution was adopted in January 2004; presidential elections, in which Karzai was elected, were held in October 2004; and National Assembly elections were held in September 2005. See the *Agreement on Provisional Arrangements in Afghanistan Pending the Re-Establishment of Permanent Government Institutions,* Bonn, December 5, 2001, available at http://www.mfa.gov.af/Documents/ImportantDoc/The%20Bonn%20Agreement.pdf.

[15] See S/RES/1386 (2001), December 20, 2001. The UK was followed by Turkey, and then Germany, see S/RES/1413 (2002), May 23, 2002, and S/RES/1444 (2002), November 27, 2002.

[16] It is a fundamental principle of military theory that war is driven by political goals of one kind or another. The Prussian writer Carl von Clausewitz argued that policy "...will permeate all military operations, and, in so far as their violent nature will admit, it will have a continuous influence on them." Carl von Clausewitz, *On War,* translated by Michael Howard and Peter Paret (Princeton, NJ: Princeton University Press, 1976.)

[17] "Strategic Vision", NATO, available at http://www.nato.int/docu/pr/2008/p08-052e.html.

[18] ISAF officials, Interviews, Kabul, Afghanistan, November 2008.

[19] National Security Council, *National Strategy for Victory in Iraq,* November 2005, available at http://www.whitehouse.gov/infocus/iraq/iraq_strategy_nov2005.html.

[20] See replies to questions for the record, submitted by Under Secretary of Defense for Policy nominee to the Senate Armed Services Committee (SASC), for her January 15, 2009, confirmation hearing, available at http://armed-services.senate.gov/statemnt/2009/January/Flournoy%2001-15-09.pdf.

[21] See the replies to questions for the record, submitted by Secretary of State nominee Hillary Clinton to the Senate Foreign Relations Committee (SFRC), for her January 13, 2009, confirmation hearing, available at http://www.foreignpolicy.com/files/KerryClintonQFRs.pdf.

[22] GEN David McKiernan, Interview, Kabul, Afghanistan, November 2008.

[23] See *The Afghanistan Compact: Building on Success*, London Conference on Afghanistan, London, January 31-February 1, 2006, available at http://www.nato.int/isaf/docu/epub/pdf/afghanistan_compact.pdf.

[24] See S/RES/1401 (2002), March 28, 2002. The mandate is renewed annually.

[25] See S/RES/1806 (2008), March 20, 2008, which extended the mandate of UNAMA for one year.

[26] Kai Eide, Interview, Kabul, Afghanistan, November 2008.

[27] Interviews with UNAMA and ISAF officials, Kabul, Afghanistan, November 2008.

[28] For further background, including the perspectives of key ISAF troop contributors, see CRS Report RL33627, *NATO in Afghanistan: A Test of the Transatlantic Alliance*, by Vincent Morelli and Paul Belkin.

[29] See Article 5, *The North Atlantic Treaty,* signed April 4, 1949, Washington, DC, available at http://www.nato.int/docu/basictxt/treaty.htm.

[30] See International Security Assistance Force "Placemat", dated December 1, 2008, available at http://www.nato.int/isaf/docu/epub/pdf/isaf_placemat_081201.pdf.

[31] General David McKiernan, Atlantic Council, Washington, DC, November 18, 2008, transcript available at http://www.acus.org/event_blog/general-david-d-mckiernan-speaks-councils-commanders-series/transcript.

[32] Interviews with ISAF officials, Kabul, Afghanistan, November 2008.

[33] Interviews with ISAF officials, Kabul, Afghanistan, November 2008.

[34] International Security Assistance Force "Placemat" dated January 12, 2009, available at http://www.nato.int/isaf/docu/epub/pdf/isaf_placemat_090112.pdf. The Placemat, which provides troop numbers for each troop-contributing country, shows two countries with "0" troops in Afghanistan; ISAF officials explained that these countries are still declared members of the coalition but had no troops on the ground on January 12th.

[35] Interview, ISAF senior official, November 2008.

[36] Pledges included 700 additional troops from France, 500 from Georgia, 400 from Poland, and 120 from the Czech Republic, as well as additional training teams from Italy, Romania and Greece. In the wake of the Russian invasion of Georgia in August 2008, Georgia has been unable to honor its pledge. See Richard Wolf, "Troop Pledges by Allies Fall Short," *USA Today,* April 23, 2008, and Steven Lee Myers and Thom Shanker, "Pentagon Considers Adding Forces in Afghanistan to Make Up for NATO Shortfall," *The New York Times*, May 3, 2008.

[37] Michael Evans, "SAS to take on Taliban 'decapitation' mission after leaving Iraq", *London Times,* December 13, 2008; and John F. Burns, "Britain Adds 300 Soldiers to Support Afghan Force," *The New York Times,* December 16, 2008.

[38] See Richard Norton-Taylor, "Hutton Attacks European Allies Over Lack of Support in Afghanistan," *The Guardian*, January 15, 2009; and Michael Smith, "Defence Chiefs Plan to Deploy 3,000 More Troops in Helmand," The Times, December 21, 2008.

[39] Minister of Defense of Afghanistan Abdul Rahim Wardak, Interview, Kabul, Afghanistan, November 2008.

[40] GEN David McKiernan and other U.S. officials, Interviews, Kabul, Afghanistan, November 2008. One additional consequence of national caveats is a tendency for U.S. troops in Afghanistan to regard ISAF with a degree of humorous skepticism – "ISAF", the line goes, stands for "I Stop At Four", or alternatively, "I Saw Americans Fighting".

[41] Major General de Kruif, Interview, Kandahar, Afghanistan, November 2008.

[42] GEN David McKiernan, Interview, Kabul, Afghanistan, November 2008.

[43] PRTs are variously civilian- or military-led, and may include any combination of civilian and military personnel, see below.

[44] Interviews with ISAF officials, Kabul, Afghanistan, November 2008. The March 2008 Report of the UN Secretary-General stressed the role of UNAMA in addressing "how to harmonize the activities of the provincial reconstruction teams." See A/62/722-S/2008/159, *The situation in Afghanistan and its implications for international peace and security, Report of the Secretary-General*, March 6, 2008.

[45] Ambassador Fernando Gentilini, Interview, Kabul, Afghanistan, November 2008.

[46] ISAF officials, Interviews, Kabul, Afghanistan, November 2008.

[47] See Matthew Cox, "10th Mountain Brigade Extended in Afghanistan," *Army Times,* January 25, 2007.

[48] See Ann Scott Tyson, "3,200 Marines to Deploy to Afghanistan in Spring," *Washington Post*, January 16, 2008.

[49] GEN David McKiernan, Atlantic Council, Washington, DC, November 18, 2008, transcript available at http://www.acus.org/event_blog/general-david-d-mckiernan-speaks-councils-commanders-series/transcript.

[50] Interviews with ISAF officials, Kabul and Kandahar, Afghanistan, November 2008.

[51] Secretary of Defense Robert Gates, Media Availability with Secretary Gates From Kandahar, Afghanistan, DoD News Transcript, December 11, 2008, available at http://www.defenselink.mil/transcripts/transcript.aspx?transcriptid=4331.

[52] ISAF officials, Interviews, Kabul and Kandahar, November 2008.

[53] GEN David McKiernan, Interview, Kabul, Afghanistan, November 2008.

[54] Gordon Lubold, "NATO Extends Afghanistan Tours," *Christian Science Monitor*, May 22,2008.

[55] GEN David McKiernan, Interview, Kabul, Afghanistan, November 2008.

[56] See Nancy A. Youssef, "U.S. Marines Find Iraq Tactics Don't Work in Afghanistan," McClatchy Newspapers, January 11, 2009. For background on MRAPs, see CRS Report RS22707, *Mine-Resistant, Ambush-Protected (MRAP)*
Vehicles: Background and Issues for Congress, by Andrew Feickert.

[57] TF Currahee and ISAF officials, Interviews, Khowst province and Kabul, Afghanistan, November 2008. Under GEN McKiernan's predecessor, ISAF went from two to four Predator lines.

[58] Christopher D. Kolenda, "How to Win in Afghanistan," *Weekly Standard*, October 13, 2008.

[59] See Secretary of Defense Robert Gates, Remarks, Air War College, Maxwell Air Force Base, Alabama, April 21, 2008, available at http://www.defenselink.mil/speeches/speech.aspx?speechid=1231. Secretary Gates said: "My concern is that our services are still not moving aggressively in wartime to provide resources needed now on the battlefield. I've been wrestling for months to get more intelligence, surveillance, and reconnaissance assets into the theatre. ...While we've doubled this capability in recent months, it is still not good enough."

[60] See "Diplomatic Note No.202," Embassy of the United States of America, Kabul, Afghanistan, September 26, 2002; "Note, Document No.791", Transitional Islamic State of Afghanistan, Ministry of Foreign Affairs, Fifth Political Department, December 12, 2002; and "Note, Document No.93", Transitional Islamic State of Afghanistan, Ministry of Foreign Affairs, American and Canada Political Affairs Department, May 28, 2003. See also Karen DeYoung, "Only a Two-Page 'Note' Governs U.S. Military in Afghanistan," *Washington Post*, August 28, 2008.

[61] See Joint Declaration of the United States-Afghanistan Strategic Partnership, May 23, 2005, available at available at http://www.whitehouse.gov/news/releases/2005/05/20050523-2.html.

[62] S/RES/1386 (2001), December 20, 2001.

[63] See Annex I, "International Security Force," *Agreement on Provisional Arrangements in Afghanistan Pending the Re-Establishment of Permanent Government Institutions*, Bonn, Germany, December 5, 2001, available at http://www.mfa.gov.af/Documents/ImportantDoc/The%20Bonn%20Agreement.pdf. The Agreement states: "This force will assist in the maintenance of security for Kabul and its surrounding areas. Such a force could, as appropriate, be progressively expanded to other urban centers and areas."

[64] S/RES/1510 (2003), October 13, 2003.

[65] UNSCR 1833 (2008), Sept 22, 2008.

[66] *The Afghanistan Compact: Building on Success*, London Conference on Afghanistan, London, UK, February 1, 2006, available at http://www.nato.int/isaf/docu/epub/pdf/afghanistan_compact.pdf.

[67] See Karen DeYoung, "Only a Two-Page 'Note' Governs U.S. Military in Afghanistan," *Washington Post*, August 28, 2008.

[68] See Associated Press, "Afghanistan Seeks More Control of NATO Troops," *Los Angeles Times*, January 21, 2009.

[69] Minister of Defense of Afghanistan Abdul Rahim Wardak, Interview, Kabul, Afghanistan, November 2008.

[70] "Metrics Brief 2007-2008," International Security Assistance Force, January 2009. The range for Afghan civilian deaths reflects differences between ISAF and UNAMA data collection. UNAMA figures reflect a lower percentage increase, but higher absolute numbers of deaths. ISAF explains that UNAMA's capacity to investigate and verify reports, *inter alia* to prevent duplication, is more limited than ISAF's, ISAF official, personal communication, January 2009.

[71] See James Kitfield, "Progress in Afghanistan Gets Rockier," *National Journal,* September 15, 2008.

[72] See Yochi Dreazen, "Taliban Expanding Foothold in Afghanistan, Report Finds," *Wall Street Journal,* December 8, 2008.

[73] Lakhdar Brahimi, "A New Path for Afghanistan," *Washington Post,* December 7, 2008.

[74] Nir Rosen, "How We Lost the War We Won," *Rolling Stone*, October 30, 2008. Rosen appeared to reach the conclusion that the Taliban was winning in part because they told him so; for example, one of his hosts noted, "From now on, it's all Taliban territory – the Americans and police don't come here at night." Of course, the same methodological questions, concerning reliance on the statements of one's informants, could also be applied to field interviews with "friendly" forces. Rosen's article also prompted some debate concerning the ethics of embedding with insurgent forces. See also Paul Watson, "Behind the Lines with the Taliban," *Los Angeles Times,* January 11, 2009. Watson, who briefly embedded with Taliban fighters in Ghazni province south of Kabul, concluded that the area was"Taliban country" where the Taliban were in charge. He quotes one fighter as saying, "Police and soldiers can never come to our territory. If they do, they won't go back safe and sound."

[75] GEN David McKiernan and other ISAF officials, Interviews, Kabul, Afghanistan, November 2008.

[76] See GEN David McKiernan, Atlantic Council, Washington, DC, November 18, 2008, transcript available at http://www.acus.org/event_blog/general-david-d-mckiernan-speaks-councils-commanders-series/transcript; and MG Robert Cone, DoD Press Briefing, November 12, 2008, available at http://www.defenselink.mil/transcripts/transcript.aspx?transcriptid=4314. Also, U.S. commanders, Interviews, November 2008.

[77] ISAF officials, Interviews, Kabul, Afghanistan, November 2008. GEN McKiernan added that overall, the Taliban is "less than the sum of its parts."

[78] UN SRSG Kai Eide, Interview, Kabul, Afghanistan, November 2008.

[79] U.S. military official, Interview, Kabul, Afghanistan, November 2008.

[80] For background about insurgent groups in Afghanistan, see Seth G. Jones, *Counterinsurgency in Afghanistan*, (Santa Monica, CA: RAND Corporation, 2008).

[81] On the Taliban in general, see Major Shahid Afsar, Pakistan Army, Major Chris Samples, U.S. Army, and Major Thomas Wood, U.S. Army, "The Taliban: An Organizational Analysis," *Military Review,* vo. 88, no. 3 (May-June 2008).

[82] Reuters, "Taliban's Murderous Mullah Threatens West", *New York Post*, December 8, 2008.

[83] See Mohammad Masoom Stanekzai, "Thwarting Afghanistan's Insurgency: A Pragmatic Approach toward Peace and Reconciliation", United States Institute of Peace, September 2008. Stanekzai, who held a fellowship at the U.S. Institute of Peace, was previously a senior GIRoA official.

[84] See Seth G. Jones, *Counterinsurgency in Afghanistan,* (Santa Monica, CA: RAND Corporation, 2008).

[85] For background about Haqqani, see Jay Solomon, "Troubled Border: Failed Courtship of Warlord Trips up U.S. in Afghanistan," *The Wall Street Journal,* November 8, 2007.

[86] ISAF senior officials, Interviews, Kabul, Afghanistan, November 2008. See Anna Mulrine, "Afghan Warlords, Formerly Backed by the CIA, Now Turn their Guns on U.S. Troops," *U.S. News and World Report,* July 11, 2008.

[87] For a concise discussion of Afghan insurgent groups and foreign groups supporting the insurgency, see Mohammad Masoom Stanekzai, "Thwarting Afghanistan's Insurgency: A Pragmatic Approach toward Peace and Reconciliation", United States Institute of Peace, September 2008.

[88] See Jane Perlez, "Taliban Leader Flaunts Power Inside Pakistan," *The New York Times,* June 2, 2008; "Tribal tribulations: The Pakistani Taliban in Waziristan," *Jane's Intelligence Review,* January 13, 2009; and "Tehrik-e-Taliban Pakistan", *Jane's Terrorism and Insurgency Centre*, January 13, 2009.

[89] See Rahimullah Yusufzai, "The emergence of the Pakistani Taliban," *Jane's Terrorism and Insurgency Centre*, December 11, 2007.

[90] James Kitfield, "Progress in Afghanistan gets rockier," *National Journal,* September 15, 2008. See also Sami Yousafzai and Ron Moreau, "The Taliban's Baghdad Strategy," *Newsweek,* August 4, 2008. The article cites a

"senior Taliban intelligence operative in Pakistan" as stating that by focusing on Kabul "we can create panic and undermine the last vestiges of support for the regime."

[91] GEN David Petraeus, U.S. CENTCOM Commanding General, recently stated, "Every case is unique. ...While general concepts that proved important in Iraq may be applicable to Afghanistan....the application of those 'big ideas ' has to be adapted to Afghanistan." GEN David Petraeus, Interview, *Foreign Policy*, (January/ February 2009), available at http://www.foreignpolicy.com/story/cms.php?story_id=4587.

[92] See David Kilcullen, "Road-Building in Afghanistan: Part 1 of a Series on Political Maneuver in Counterinsurgency," *Small Wars Journal,* April 24, 2008, available at http://smallwarsjournal.com/blog/2008/04/political-maneuver-in-counteri/; and David Ignatius, "Building Bridges in the Back of Beyond," *Washington Post,* May 1, 2008.

[93] 24th MEU officials, Interviews, Kandahar, Afghanistan, November 2008.

[94] See for example GEN David Petraeus, COIN Guidance, Multi-National Force Iraq, 2008.

[95] RC-East officials, Interviews, Bagram, Afghanistan, November 2008.

[96] RC-East official, Interview, Bagram, Afghanistan, November 2008.

[97] H.R. McMaster, "The Human Element: When Gadgetry Becomes Strategy," *World Affairs*, Winter 2009.

[98] For a recent analysis of the kinetic use of air in Afghanistan, which argues against "excessive restraint in the use of airstrikes", see Lara M. Dadkhah, "Close Air Support and Civilian Casualties in Afghanistan," *Small Wars Journal*, December 30, 2008, available at http://smallwarsjournal.com/blog/2008/12/close-air-support-and-civilian/.

[99] See James Kitfield, "Progress in Afghanistan Gets Rockier," *National Journal,* September 15, 2008; and Carlotta Gall, "Afghan Leader Assails Airstrike He Says Killed 95," *The New York Times,* August 23, 2008.

[100] Candace Rondeaux and Karen DeYoung, "U.N. Finds Airstrike Killed 90 Afghans," *Washington Post*, August 27, 2008.

[101] Kai Eide, Interview, Kabul, Afghanistan, November 2008.

[102] Kirk Semple, "Official Calls for Sensitivity to Afghan Demands," *The New York Times*, December 8, 2008.

[103] ISAF slide, 'Tactical Directive", October 2, 2008.

[104] GEN David McKiernan, Interview, Kabul, Afghanistan, November 2008.

[105] GEN David Petraeus, Interview, Foreign Policy, January/ February 2009, available at http://www.foreignpolicy.com/story/cms.php?story_id=4587.

[106] For a detailed discussion, see CRS Report RL33498, *Pakistan-U.S. Relations,* by K. Alan Kronstadt.

[107] For background and analysis of the FATA, see Daniel Markey, "Securing Pakistan's Tribal Belt," Council on Foreign Relations, *Council Special Report* no. 36, August 2008.

[108] See Barnett R. Rubin and Ahmed Rashid, "From Great Game to Grand Bargain: Ending Chaos in Afghanistan and Pakistan," *Foreign Affairs*, vol. 87, no. 6 (November/ December 2008). The authors add, "The area is kept underdeveloped and over-armed as a barrier against invaders".

[109] According to unclassified CIA estimates from July 2008, ethnic Pashtuns comprise 15.42% of Pakistan's population of 172,800,048; and about 42% of Afghanistan's population of 32,738,376. See "Pakistan" and "Afghanistan", The World Fact Book, Central Intelligence Agency, available at https://www.cia.gov/library/publications/the-world-factbook/geos/pk.html and https://www.cia.gov/library/publications/the-world-factbook/geos/af.html.

[110] Text of agreement available at http://www.pbs.org/wgbh/pages/frontline/taliban/etc/nwdeal.html.

[111] See Jane Perlez, "Taliban Leader Flaunts Power Inside Pakistan," *The New York Times*, June 2, 2008, and Jon Hemming, "NATO Beefs Up Forces Along Afghan-Pakistan Border," *Reuters,* May 19, 2008. See also Daniel Markey, "Securing Pakistan's Tribal Belt," Council on Foreign Relations, *Council Special Report* no. 36, August 2008, p. 11.

[112] See Karen DeYoung, "Pakistan Will Give Arms to Tribal Militias," *Washington Post,* October 23, 2008, and Saeed Shah, "Pakistan Rejects 'America's War' on Extremists, *The Guardian*, October 24, 2008.

[113] ISAF, Interviews, Kabul, Afghanistan, November 2008.

[114] Hamid Shalizi, "NATO Chief Says Will Not Hunt Taliban in Pakistan," *Reuters,* July 24, 2008.

[115] U.S. military officials, Interviews, Kabul and Bagram, Afghanistan, November 2008. More than one official cited in this context the phrase from Roman historian Tacitus: "They make a desert and call it peace."

[116] See Karen DeYoung, "Pakistan Will Give Arms to Tribal Militias," *Washington Post*, October 23, 2008. Such approaches are not new. Just before independence, in 1946, the British government issued a statement noting a reconsideration of its frontier policy. They would now "enlist cooperative tribesmen themselves", rather than simply bombing the area. See "British End Bombing of Indian Tribesmen," *The New York Times,* September 13, 1946.

[117] GEN David McKiernan, Interview, Kabul, Afghanistan, November 2008.

[118] Major General Jeffrey Schloesser, Interview, Bagram, Afghanistan, November 2008.

[119] Task Force Currahee officials, Interviews, Khowst Province, Afghanistan, November 2008.

[120] GEN David McKiernan, Atlantic Council, Washington, DC, November 18, 2008, transcript available at http://www.acus.org/event_blog/general-david-d-mckiernan-speaks-councils-commanders-series/transcript.

[121] U.S. military officials, Interviews, Kabul and Bagram, Afghanistan, November 2008

[122] Saeed Shah, "Pakistan Rejects 'America's War' on Extremists," *The Guardian*, October 24, 2008. This resolution was passed before the late November 2008 terrorist attacks in Mumbai.

[123] On Iran's behavior at the Bonn Conference, see Ambassador James F. Dobbins, *After the Taliban: Nation-Building in Afghanistan*, (Washington, DC: Potomac Books, 2008).

[124] ISAF and U.S. officials, Interviews, Kabul, Bagram, and Kandahar, Afghanistan, November 2008.

[125] *The National Defense Authorization Act for Fiscal Year 2008*, P.L. 110-181, January 28, 2008, §1230, required the Department of Defense to submit a "report on progress toward security and stability in Afghanistan", and §1231 required a report on sustaining the Afghanistan National Security Forces. DoD submitted the first edition of both reports in June 2008. See Department of Defense, *Report on Progress Toward Security and Stability in Afghanistan,* June 2008, available at http://www.defenselink.mil/pubs/Report_on_Progress_toward_Security_and_Stability_in_Afghanistan_1230.p df; and Department of Defense, *United States Plan for Sustaining the Afghanistan National Security Forces,* June 2008, available at http://www.defenselink.mil/pubs/united_states_plan_for_sustaining_the_afghanistan_national_security_forces _1231.p df.

[126] "Command Changes Hands for Afghan Security Forces Training," *American Forces Press Service*, December 22, 2008; and information from ISAF, December 2008.

[127] CSTC-A officials, Interviews, Kabul, Afghanistan, November 2008.

[128] As of the end of 2008, most ANA recruits received most of their initial training at the KMTC. Significant construction was underway to improve the KMTC physical plant, but doubling the size of each class of recruits, for example, would require significantly more space in buildings and on ranges. The Basic Warrior Training course was 10 weeks long, and a new class of 1,200 – the size of an ANA "kandak" (battalion) – was admitted every two weeks. As of November 2008, the KMTC had 3,000 Afghan personnel, of whom 1,200 were trainers, and the near-term goal was to increase the number of Afghan trainers by 300. The "122 plus 12" target number for the ANA does not include additional Afghan trainers, and Afghan trainers are one of the few groups in the ANA that does not receive additional combat or incentive pay of some kind. KMTC senior officials, and U.S. and Afghan staff, Interviews, Kabul, Afghanistan, November 2008.

[129] GEN David McKiernan, MG Robert Cone, U.S. military officials, Interviews, Kabul, Kandahar, and Bagram, Afghanistan, November 2008.

[130] Minister of Defense of Afghanistan Abdul Rahim Wardak, Interview, Kabul, Afghanistan, November 2008.

[131] Mohammad Masoom Stanekzai, "Thwarting Afghanistan's Insurgency: A Pragmatic Approach toward Peace and Reconciliation", United States Institute of Peace, September 2008.

[132] "All Things Considered," National Public Radio, November 18, 2008, interview with John Nagl, transcript available at http://www.npr.org/templates/story/story.php?storyId=97170621: ANA should grow to 250,000.

[133] ANA Corps Commanding Generals, Interviews, Kandahar and Mazar-e-Sharif, Afghanistan, November 2008.

[134] Minister of Defense of Afghanistan Abdul Rahim Wardak, Interview, Kabul, Afghanistan, November 2008.

[135] See http://www.mod.gov.af, which calls itself the "website of the Afghan National Army."

[136] "Metrics Brief 2007-2008," International Security Assistance Force, January 2009.

[137] CSTC-A officials, Interviews, Kabul, Afghanistan, November 2008.

[138] ISAF officials, Interviews, Kabul, Bagram, and Khowst Province, Afghanistan, November 2008.

[139] CSTC-A officials, Interviews, Kabul, Afghanistan, November 2008; and see Gordon Lubold, "Americans Build Elite Afghan Commando Force," *Christian Science Monitor,* May 1, 2008.

[140] MG Robert Cone, other CSTC-A officials, Interviews, Kabul, Afghanistan, November 2008.

[141] Minister of Defense of Afghanistan Abdul Rahim Wardak, Interview, Kabul, Afghanistan, November 2008.

[142] *Afghanistan in 2008: A Survey of the Afghan People*, The Asia Foundation, 2008, p.25, available at http://asiafoundation.org/country/afghanistan/2008-poll.php. Other public institutions considered included the media, NGOs, national, provincial and local governing bodies, and community organizations. In the survey, 89% of respondents agreed that the ANA is "honest and fair with the Afghan people" (48% strongly agreed, 41% somewhat agreed), and 89% of respondents agreed that the "ANA helps improve security" (51% strongly agreed, and 35% somewhat agreed). This apparent high regard was not unqualified; 55% of respondents agreed that the "ANA is unprofessional and poorly trained" (18% strongly agreed, and 37% somewhat agreed), see p.35.

[143] CAPTF officials, Interviews, Kabul, Afghanistan, November 2008.

[144] CAPTF officials, Interviews, Kabul, Afghanistan, November 2008.

[145] CAPTF officials, Interviews, Kabul, Afghanistan, November 2008.

[146] CAPTF officials, Interviews, Kabul, Afghanistan, November 2008.

[147] See Department of Defense, United States Plan for Sustaining the Afghanistan National Security Forces, June 2008, p.21, available at http://www.defenselink.mil/pubs/united_states_plan_for_sustaining_the_afghanistan _national_security_forces_1231.pdf.

[148] "Command Changes Hands for Afghan Security Forces Training," *American Forces Press Service*, December 22, 2008.

[149] GEN David McKiernan, Atlantic Council, Washington, DC, November 18, 2008, transcript available at http://www.acus.org/event_blog/general-david-d-mckiernan-speaks-councils-commanders-series/transcript. ISAF officials, Interviews, Kabul, Kandahar, Bagram, Afghanistan, November 2008.

[150] See Sarah Chayes, "Clean Up the Afghan Government, and the Taliban Will Fade Away," *Washington Post*, December 14, 2008. The author lives in Kandahar and runs an Afghan cooperative. She argues: "Now, Afghans are suffering so acutely that they hardly feel the difference between Taliban depredations and those of their own government." She quotes one local resident of Kandahar as saying: "The Taliban shake us down at night, and the government shakes us down in the daytime."

[151] *Afghanistan in 2008 - A Survey of the Afghan People*, The Asia Foundation, 2008, p.25, available at http://asiafoundation.org/country/afghanistan/2008-poll.php. In the survey, 80% of respondents agreed that the "ANP is honest and fair with the Afghan people," (40% strongly, 40% somewhat); while 80% agreed that the "ANP helps improve security" (40% strongly, 40% somewhat). At the same time, 60% agreed that the "ANP is unprofessional and poorly trained" (22% strongly, 38% somewhat).

[152] "Metrics Brief 2007-2008," International Security Assistance Force, January 2009.

[153] International organization official, Interview, Kabul, Afghanistan, November 2008.

[154] Coalition officials, Interviews, Kabul, Afghanistan, November 2008.

[155] "CJTF-101 Campaign Update" slides, November 2008.

[156] MG Robert Cone, Interview, Kabul, Afghanistan, November 2008.

[157] Minister of Defense of Afghanistan Abdul Rahim Wardak, Interview, Kabul, Afghanistan, November 2008.

[158] The inter-service coordination arrangements in Afghanistan differ from those in Iraq, also designed for the exigencies of a counterinsurgency effort, where provincially-based "Operations Commands" bring together multiple Iraqi security forces under the formal command of the head of each Operations Command. Those arrangements may prove more difficult to rationalize, for a future post-COIN environment, than the OCC's in Afghanistan.

[159] U.S. Government funding support to the ANSF has not followed a smooth trajectory to date. In Fiscal Year 2006, Congress appropriated approximately $1.9 billion for the Afghan Security Forces Fund, and in Fiscal Year 2007 – meeting DoD's request based on the need to accelerate Afghan ground forces training and equipping – $7.4 billion. In Fiscal Year 2008, Congress appropriated approximately $2.7 billion for the Fund, again meeting DoD's request, which was based on the premise of building on the ANSF acceleration enabled by the FY 2007 appropriations. For Fiscal Year 2009, DoD requested an increase – $3.7 billion – in its Global War on Terror bridge request, submitted in May 2008, of which Congress provided $2 billion. See CRS Report RL33110, *The Cost of Iraq, Afghanistan, and Other Global War on Terror Operations Since 9/11*, by Amy Belasco, p.22. See also Department of Defense, "FY 2007 Emergency Supplemental Request for the Global War on Terror," February 2007, available at http://www.defenselink.mil/comptroller/Docs/FY2007_ Emergency_Supplemental_Request_for_the_GWOT.pdf; Department of Defense, "FY 2008 Global War on Terror Request," February 2007, available athttp://www.defenselink.mil/comptroller/Docs/FY2008_February_ Global_War_On_Terror_Request.pdf; and Department of Defense, "Fiscal Year 2009 Global War on Terror Bridge Request," May 2008, available at http://www.defenselink.mil/comptroller/defbudget/fy2009/ Supplemental/FY2009_Global_War_On_Terror_Bridge_Re quest.pdf.

[160] See the replies to questions for the record, submitted by Secretary of State nominee Hillary Clinton to the Senate Foreign Relations Committee (SFRC), for her January 13, 2009, confirmation hearing, available at http://www.foreignpolicy.com/files/KerryClintonQFRs.pdf.

[161] In a 2008 report, regarding the establishment of the OSC-A, DoD noted: "Efforts prior to this time were not comprehensive and lacked both resources and unity of effort within the international community". See Department of Defense, *United States Plan for Sustaining the Afghanistan National Security Forces*, June 2008, p.21, available at http://www.defenselink.mil/pubs/united_states_plan_for_sustaining_the_afghanistan _national_security_forces_1231.p df. The OSC-A, like the OMC-A, reported to the U.S. military command in Afghanistan, but received policy guidance from the U.S. Chief of Mission, while contract management authority remained with State INL.

[162] ISAF and CSTC-A officials, Interviews, Kabul, Kandahar, Bagram, Afghanistan, November 2008.

[163] John Milburn, "Adviser Mission to Leave Fort Riley," *Army Times,* October 1, 2008. The U.S. Marine Corps has its own program for preparing Marines for advisory roles.

[164] MG Robert Cone, Interview, Kabul, Afghanistan, November 2008.

[165] MG Robert Cone, DoD Press Briefing, November 12, 2008, available at http://www.defenselink.mil/transcripts/transcript.aspx?transcriptid=4314.

[166] John F. Burns, "Afghan President, Pressured, Reshuffles Cabinet," *The New York Times*, October 11, 2008.

[167] Task Force Currahee, Interviews, Khowst Province, Afghanistan, November 2008.

[168] Minister of Defense of Afghanistan Abdul Rahim Wardak, Interview, Kabul, Afghanistan, November 2008.

[169] CSTC-A officials, Interviews, Kabul, Afghanistan, November 2008.

[170] Available at http://consilium.europa.eu/cms3_fo/showPage.asp?id=1268&lang=EN. See also Council Joint Action 2007/369/CFSP, May 30, 2007, on the establishment of the European Union Police Mission in Afghanistan, available at http://eurlex.europa.eu/LexUriServ/site/en/oj/2007/l_139/

l_13920070531en00330038.pdf. EUPOL objectives, listed in Article 3, include contributing to "sustainable and effective civilian policing arrangements" and supporting "the reform process towards a trusted and efficient police service."

[171] Minister of Defense of Afghanistan Abdul Rahim Wardak, MG Robert Cone, Interviews, Kabul, Afghanistan, November 2008.

[172] Kai Eide, Interview, Kabul, Afghanistan, November 2008.

[173] GEN David McKiernan, Atlantic Council, Washington, DC, November 18, 2008, transcript available at http://www.acus.org/event_blog/general-david-d-mckiernan-speaks-councils-commanders-series/transcript; and "U.S. Backs Plan for Engaging Afghan Tribes," *Radio Free Europe Radio Liberty*, December 30, 2008.

[174] Minister of Defense of Afghanistan Abdul Rahim Wardak, Interview, Kabul, Afghanistan, November 2008.

[175] ISAF, RC-South officials, and western diplomats, Interviews, Kabul, Afghanistan, November 2008.

[176] Jon Boone, "Afghans Fear U.S. Plan to Rearm Villages," *Financial Times*, January 12, 2009.

[177] ISAF officials, Interviews, Kabul, Afghanistan, November 2008.

[178] Ambassador William Wood, Interview, Kabul, Afghanistan, November 2008.

[179] Minister of Defense of Afghanistan Abdul Rahim Wardak, Interview, Kabul, Afghanistan, November 2008. He added, "If you make a community responsible, and you given them the feeling that they are sharing in that responsibility, then they will die for it."

[180] RC-East official, Interview, Bagram, Afghanistan, November 2008.

[181] Minister of Defense of Afghanistan Abdul Rahim Wardak, Interview, Kabul, Afghanistan, November 2008.

[182] RC-East official, Interview, Bagram, Afghanistan, November 2008.

[183] See Andrew Wilder, "Cops or Robbers? The Struggle to Reform the Afghan National Police", *Afghanistan Research and Evaluation Unit,* July 2007.

[184] See Jerome Starkey, "US attacks UK plan to arm Afghan militias", *The Independent*, January 14, 2008; and Murray Brewster, "NATO Disbands Afghan Auxiliary Police," *Edmonton Sun*, May 15, 2008; MG Robert Cone, Interview, Kabul, Afghanistan, November 2008; RC-East official, Interview, Bagram, Afghanistan, November 2008.

[185] GEN David McKiernan, Interview, Kabul, Afghanistan, November 2008. See also Genral David McKiernan, Atlantic Council, Washington, DC, November 18, 2008, transcript available at http://www.acus.org/event_blog/general-david-d-mckiernan-speaks-councils-commanders-series/transcript.

[186] See for example Anand Gopal, "Bid to Split Taliban, Al Qaeda," *The Christian Science Monitor,* December 16, 2008.

[187] See John F. Burns, "Karzai Sought Saudi Help With Taliban," *The New York Times*, October 1, 2008; and Anand Gopal, "No Afghan-Taliban Peace Talks for Now," *Christian Science Monitor*, October 9, 2008.

[188] Kim Sengupta, "Secret Saudi Dinner, Karzai's Brother and the Taliban," *The Independent,* October 8, 2008.

[189] Senior U.S. officials, Interviews, Kabul, Afghanistan, November 2008; and see John F. Burns, "Karzai Sought Saudi Help With Taliban," *The New York Times*, October 1, 2008.

[190] GEN David McKiernan, Interview, Kabul, Afghanistan, November 2008.

[191] Kai Eide, Interview, Kabul, Afghanistan, November 2008.

[192] UK official, Interview, Kabul, Afghanistan, November 2008.

[193] ISAF officials, Interviews, Kabul, Afghanistan, November 2008.

[194] For a detailed discussion, see CRS Report RL32686, *Afghanistan: Narcotics and U.S. Policy*, by Christopher M. Blanchard.

[195] *Afghanistan Opium Survey 2008*, United Nations Office on Drugs and Crime, November 2008.

[196] Government of Afghanistan, *Afghan National Drug Control Strategy: An Updated Five-Year Strategy for Tackling the Illicit Drug Problem,* Islamic Republic of Afghanistan Ministry of Counter-Narcotics, Kabul, January 2006.

[197] Ambassador Thomas A. Schweich, Coordinator for Counternarcotics and Justice Reform in Afghanistan, *U.S. Counternarcotics Strategy for Afghanistan,* August 2007, see especially pp.5-6, and 48.

[198] NATO press release, October 10, 2008, available at http://www.nato.int/docu/update/2008/10-october/e1010b.html.

[199] GEN David McKiernan, Atlantic Council, Washington, DC, November 18, 2008, transcript available at http://www.acus.org/event_blog/general-david-d-mckiernan-speaks-councils-commanders-series/transcript.

[200] Dave Pugliese, "ISAF Commander Says Afghan Drug Lords to Be on 'Kill or Capture' Target List," *Ottawa Citizen*, December 23, 2008.

[201] *Afghanistan Opium Survey 2008*, United Nations Office on Drugs and Crime, November 2008, p.25.

[202] "Farm gate" cultivation was valued at $1 billion in 2007, and $730 million in 2008. 193,000 hectares of poppy were cultivated in 2007, and 157,000 hectares were cultivated in 2008, a decline of 19%. 2.5% of arable land was cultivated in 2007, while 2.1% was cultivated in 2008. In 2007, 13 provinces were considered "poppy-free", while 18 were so characterized in 2008. *Afghanistan Opium Survey 2008*, United Nations Office on Drugs and Crime, November 2008, pp.2, 5, 7.

[203] Bureau of International Narcotics and Law Enforcement Affairs, "Update on Counternarcotics in Afghanistan," U.S. Department of State, August 2008.

204 David Mansfield and Adam Pain, "Counter-Narcotics in Afghanistan: The Failure of Success?" *Afghanistan Research and Evaluation Unit Briefing Paper Series,* December 2008, p.3.

205 Clare Lockhart, "Learning from Experience," *Slate*, November 5, 2008.

206 The 2006 COIN field manual states: "Political, social and economic programs are most commonly and appropriately associated with civilian organizations and expertise; however, effective implementation of these programs is more important than who performs the tasks. If adequate civilian capacity is not available, military forces fill the gap." Field Manual 3-24, *Counterinsurgency,* Headquarters, Department of the Army, December 2006, para. 2-5.

207 ISAF official, Interview, Kabul, Afghanistan, November 2008.

208 Ambassador Ferrnando Gentilini, Interview, Kabul, Afghanistan, November 2008.

209 ISAF official, Interview, Kandahar, Afghanistan, November 2008.

210 See Independent Directorate of Local Governance (IDLG), "Afghanistan Social Outreach Programme: Programme Document," Islamic Republic of Afghanistan, April 2008.

211 The IDLG convened a Governors Meeting for the southern region of Afghanistan in Kandahar in August 2008, and in Oruzgan in November 2008.

212 Kai Eide, Interview, Kabul, Afghanistan, November 2008.

213 See Independent Directorate of Local Governance (IDLG), "Afghanistan Social Outreach Programme: Programme Document," Islamic Republic of Afghanistan, April 2008.

214 ISAF officials, Interviews, Kabul and Bagram, Afghanistan, November 2008.

215 "CJTF-101 Campaign Update" slides, November 2008.

216 Ann Marlowe, "A Counterinsurgency Grows in Khowst," *Weekly Standard,* May 19, 2008. Colonel Marty Schweitzer led the 4th BCT, 82nd Airborne Division, in 6 provinces of RC-East.

217 ISAF official, Interview, Kabul, Afghanistan, November 2008.

218 For a current U.S. PRT mission statement, see "Fact Sheet: Making Afghanistan More Secure with Economic and Reconstruction Assistance," White House, September 26, 2008, available at http://www.whitehouse.gov/news/releases/2008/09/20080926-16.html.

219 Zabul PRT officials, Interviews, Zabul province, Afghanistan, November 2008.

220 The constraint of limited resources may also apply to U.S.-led PRTs. In late 2008, the Zabul PRT introduced a new ionizer-based water purification initiative – a scheme developed and proposed by an American Eagle Scout. While that Eagle Scout's contributions may be laudable, some suggest that PRTs ought to be able to draw on more readily available expertise.

221 Associated Press, "Afghan President Complains U.S., NATO Aren't Succeeding," Foxnews.com, November 26, 2008.

222 ISAF officials, and Minister of Defense of Afghanistan Abdul Rahim Wardak, Interviews, Kabul, Afghanistan, November 2008.

223 RC-East, TF Currahee officials, Interviews, Bagram and Khowst province, Afghanistan, November 2008.

224 Zabul PRT officials, Interviews, Zabul province, Afghanistan, November 2008.

225 For information, see the Human Terrain System website, http://humanterrainsystem.army.mil/default.htm.

226 RC-East and TF Currahee officials and HTT members, Interviews, Bagram and Khowst province, Afghanistan, November 2008. To be clear, HTT members state that they "do not do targeting."

227 See for example, Network of Concerned Anthropologists, "Anthropologists Up in Arms Over Pentagon's 'Human Terrain System' to Recruit Graduate Students to Serve in Iraq, Afghanistan: an interview with David Price," *Democracy Now*, December 13, 2007, available at http://www.democracynow.org/2007/12/13/anthropologists_up_in_arms_over_pentagons.

228 See Jeff Schogol, "DoD Contractor Charged in Death of Afghan Man," *Stars and Stripes,* November 22, 2008; and Jerry Markon, "Contractor Charged," *Washington Post*, November 20, 2008.

229 RC-East and TF Currahee officials, Interviews, Bagram and Khowst province, Afghanistan, November 2008.

230 See Special Inspector General for Afghanistan Reconstruction, *Quarterly Report to the United States Congress*, October 2008, p.39

231 "CJTF-101 Campaign Update," slides, November 2008.

232 RC-East and TF Currahee officials, Interviews, Bagram and Khowst provinces, Afghanistan, November 2008. See also Tom Vanden Brook, "Harvesting Ties with Afghanistan," *Army Times*, December 31, 2008.

233 MG Jeffrey Schloesser, Interview, Bagram, Afghanistan, November 2008.

234 MG Weigt, Interview, Mazar-e-Sharif, Afghanistan, November 2008.

235 Sarah Chayes, "Clean up the Afghan Government, and the Taliban will Fade Away," *Washington Post*, December 14, 2008.

236 See for example "Afghanistan: Winning the War. Problems, Solutions, and Critical Decisions for the Next Administration," Foundation for the Defense of Democracies, Washington DC, December 4, 2008, panel discussion video transcript available at http://www.defenddemocracy.org/index.php?option=com_content&view=article&id=11784099&Itemid=385.

237 U.S. Embassy and Department of State officials, Interviews, Kabul, Afghanistan and Washington, DC, November and December 2008.

[238] *Afghanistan National Development Strategy 1387-1391 (2008-2013): A Strategy for Security, Governance, Economic Growth and Poverty Reduction*, Islamic Republic of Afghanistan, 2008, available at http://www.ands.gov.af/ands/ands_docs/index.asp.

[239] Lakhdar Brahimi, "A New Path for Afghanistan," Washington Post, December 7, 2008. The Afghanistan Study Group, which released its results in January 2008, called for developing not only "a long-term, coherent international strategy" but also "a strategic communications plan to garner strong public support for that strategy" among NATO countries and regional leaders. If "strategic communications" is regarded as two-way dialogue, then the Group's proposal is similar to Brahimi's. See General James L. Jones, USMC (ret), Ambassador Thomas R. Pickering, Co-Chairs, Afghanistan Study Group Report, "Revitalizing our Efforts, Rethinking our Strategies," Center for the Study of the Presidency, January 30, 2008, Page 11.

[240] Tom Vanden Brook, "A 'Tough Fight' Seem for Afghan War in '09", *USA Today,* December 8, 2009.

[241] Tom Vanden Brook, "A 'Tough Fight' Seem for Afghan War in '09", *USA Today*, December 8, 2009. Vanden Brook interviews GEN McKiernan, whom he quotes as saying: "I don't like to use the word surge here because if we put these additional forces in here, it's going to be for the next few years. It's not a temporary increase of combat strength."

[242] See Rory Stewart, "How to Save Afghanistan," *Time*, July 17, 2008. Stewart, the author of *The Places in Between*, is a former British diplomat and the head of the Turquoise Mountain project in Kabul.

[243] Barnett R. Rubin, "Afghan Dilemmas: Defining Commitments," *The American Interest,* vol. 3, no.5 (May/June 2008), pp.45-46.

[244] See the replies to questions for the record, submitted by Secretary of State nominee Hillary Clinton to the Senate Foreign Relations Committee (SFRC), for her January 13, 2009, confirmation hearing, available at http://www.foreignpolicy.com/files/KerryClintonQFRs.pdf.

[245] See the replies to questions for the record, submitted by Secretary of State nominee Hillary Clinton to the Senate Foreign Relations Committee (SFRC), for her January 13, 2009, confirmation hearing, available at http://www.foreignpolicy.com/files/KerryClintonQFRs.pdf.

[246] GEN David McKiernan, Interview, Kabul, Afghanistan, November 2008.

[247] ISAF official, Interview, Kabul, Afghanistan, November 2008. Recent attacks on the supply line from Karachi, Pakistan, to the Afghanistan border included the December 7, 2008, insurgent attacks on two truck stops in Peshawar, Pakistan, destroying over 100 vehicles with supplies bound for Afghanistan, and the November 10, 2008, insurgent hijacking of a convoy of vehicles on the road to the Khyber pass, bound for Afghanistan. See Laura King, "Suspected Taliban Militants Destroy Supply Trucks," *Los Angeles Times*, December 8, 2008.

[248] Chris Brummitt, "U.S. Reaches Deal on Afghan Supply Routes," *Army Times*, January 20, 2009.

[249] See Otto Kreisher, "Marine Commandant Expects Troop Surge in Afghanistan," *National Journal*, January 15, 2009. See also Gordon Lubold, "A 'Surge' for Afghanistan?, The Christian Science Monitor, November 29, 2007; Lubold cited General James Conway, Commandant of the Marine Corps, as saying: "The trend lines tell us that it may be time to increase the force posture in Afghanistan. …If it requires additional US forces, then it goes back to our suggestion that maybe we need more Marines in there with a more kinetic bent." Gen Conway has also expressed concern that the Marines have "become, in some ways, a second land army," and may be losing their "expeditionary flavor," a danger that deployment to Afghanistan might help to avert. See Remarks by General James T. Conway, Commandant of the Marine Corps, Center for a New American Security, Washington DC, October 15, 2007, Available at http://www.marines.mil/units/hqmc/cmc/Documents/Speeches20071015CNAS.pdf.

[250] See General James L. Jones, USMC (ret), Ambassador Thomas R. Pickering, *Afghanistan Study Group Report: Revitalizing our Efforts, Rethinking our Strategies,* Center for the Study of the Presidency, January 30, 2008, p.23. The authors argued for an enhanced international effort: "The U.S. and its NATO partners also need to focus more efforts and resources on training and standing up the Afghan National Army and recruiting, training, and providing adequate pay and equipment to the Afghan National Police so they can maintain security in an area once coalition forces depart."

[251] MG Robert Cone, Interview, Kabul, Afghanistan, November 2008.

[252] Minister of Defense of Afghanistan Abdul Rahim Wardak, Interview, Kabul, Afghanistan, November 2008.

[253] MG Robert Cone, Interview, Kabul, Afghanistan, November 2008.

[254] Christopher D. Kolenda, "How to Win in Afghanistan," *Weekly Standard*, October 13, 2008. Colonel Christopher Kolenda recently commanded a U.S. Army battalion in Afghanistan along the border with Pakistan.

[255] See Thomas Lynch, "Afghan Dilemmas: Staying Power," *The American Interest,* vol. 3, no.5 (May/June 2008), p.36. U.S. Army Colonel Thomas Lynch has served in Afghanistan.

[256] See Vikram J. Singh and Nathaniel C. Fick, "Surging Statecraft to Save Afghanistan", *Small Wars Journal*, October 2, 2008.

[257] Barnett R. Rubin, "Afghan Dilemmas: Defining Commitments," *The American Interest,* vol. 3, no.5 (May/June 2008), p.49. He adds that the United States "must instead support civilian control over the government and military alike, even by parties that oppose U.S. objectives openly (rather than covertly, like the military)."

[258] See the replies to questions for the record, submitted by Secretary of State nominee Hillary Clinton to the Senate Foreign Relations Committee (SFRC), for her January 13, 2009, confirmation hearing, available at http://www.foreignpolicy.com/files/KerryClintonQFRs.pdf. Under Secretary of Defense for Policy Michèle Flournoy, in replies to questions for the record submitted for her own confirmation hearing, with the Senate Armed Services Committee on January 15, 2009, echoed the same premise, stating: "The U.S. must have an integrated strategy to promote development and prevent terrorism across the Afghanistan-Pakistan border region." See replies to questions for the record, submitted by Under Secretary of Defense for Policy nominee to the Senate Armed Services Committee (SASC), for her January 15, 2009, confirmation hearing, available at http://armedservices.senate.gov/statemnt/2009/January/Flournoy%2001-15-09.pdf.

[259] See for example General James L. Jones, USMC (ret), Ambassador Thomas R. Pickering, *Afghanistan Study Group Report: Revitalizing our Efforts, Rethinking our Strategies,* Center for the Study of the Presidency, January 30, 2008, Page 17.

[260] General James L. Jones, USMC (ret), Ambassador Thomas R. Pickering, *Afghanistan Study Group Report: Revitalizing our Efforts, Rethinking our Strategies,* Center for the Study of the Presidency, January 30, 2008, Page 17.

[261] Department of Defense, "Irregular Warfare (IW)", DoDD 3000.07, December 1, 2008. The Directive defined IW this way: "A violent struggle among state and non-state actors for legitimacy and influence over the relevant population(s). Irregular warfare favors indirect and asymmetric approaches, though it may employ the full range of military and other capacities, in order to erode an adversary's power, influence, and will." It noted that "IW can include a variety of steady-state and surge DoD activities and operations: counter-terrorism, unconventional warfare; foreign internal defense; counterinsurgency; and stability operations that, in the context of IW, involve establishing or re-establishing order in a fragile state."

[262] Rory Stewart, "How to Save Afghanistan," *Time,* July 17, 2008.

[263] Clare Lockhart, "Learning from Experience," *Slate,* November 5, 2008.

CHAPTER SOURCES

The following chapters have been previously published:

Chapter 1 - This is an edited, reformatted and augmented version of a Congressional Research Service publication, Report Order Code RL33627, dated January 23, 2009.

Chapter 2 – This website information has been edited, reformatted and augmented from www.StrategicStudiesInstitute.army.mil.

Chapter 3 - This is an edited, reformatted and augmented version of a Congressional Research Service publication, Report Order Code R40156, dated January 23, 2009.

INDEX

B

C

D

E

F

G

H

I

M

N

O

T

U

V

W

Y